国际建筑师协会
关于建筑实践中职业主义的推荐国际标准

许安之 编译

中国建筑工业出版社

图书在版编目(CIP)数据

国际建筑师协会关于建筑实践中职业主义的推荐国际标准/许安之编译．
北京：中国建筑工业出版社，2004
ISBN 7-112-06847-9

Ⅰ．国… Ⅱ．许… Ⅲ．建筑设计—国际标准 Ⅳ．TU2-65

中国版本图书馆 CIP 数据核字(2004)第 087146 号

责任编辑：王莉慧
责任设计：刘向阳
责任校对：赵明霞

国际建筑师协会
关于建筑实践中职业主义的推荐国际标准
许安之 编译
*
中国建筑工业出版社 出版、发行(北京西郊百万庄)
新 华 书 店 经 销
北京建筑工业印刷厂印刷
*

开本：880×1230 毫米 1/32 印张：5¼ 字数：150 千字
2005 年 1 月第一版 2005 年 1 月第一次印刷
印数：1—2,500 册 定价：**14.00** 元
ISBN 7-112-06847-9
　TU·6094(12801)
版权所有　翻印必究
如有印装质量问题，可寄本社退换
(邮政编码 100037)
本社网址：http://www.china-abp.com.cn
网上书店：http://www.china-building.com.cn

前　言

20世纪90年代开始，经济全球化进程从货物贸易延伸到服务贸易，世界贸易组织（WTO）于1994年签订了服务贸易总协定（GATS），从此，各国建筑师跨境的专业服务被纳入WTO服务贸易的范围。

建筑师职业在古代世界就有，建筑师职业团体在一些国家也有较长的历史，如英国皇家建筑师学会（RIBA）成立于1837年。全球性的建筑师职业团体的联合会——国际建筑师协会（UIA）于第二次世界大战后不久成立，现有各国和地区的成员组织130多个，代表全球绝大多数已建立建筑师职业团体的国家和地区的建筑师。

与经济全球化的进程相适应，国际建筑师协会从1994年开始设立了职业实践委员会，由美国建筑师学会的詹姆斯·席勒和中国建筑学会的张钦楠先生共同担任委员会的主任。1999年7月后，本人接替张钦楠先生在UIA职业实践委员会的工作。在国际建筑师协会主席和秘书长的领导下，在各成员组织的共同参与下，《国际建筑师协会关于建筑师职业实践中职业主义的推荐国际标准》第一版于1996年在巴塞罗那UIA第20次代表大会上通过，经过征求各国建筑师职业团体意见并进行修改后的第二版，在1999年7月于北京举行的UIA第21届代表大会上获得一致通过。从此，各国建筑师有了一部共同认同的职业"宪章"——《国际标准认同书》简称《国际建协认同书》或《UIA认同书》。

在1994—1999年期间及后来的1999—2002年期间，国际建筑师协会还通过了一系列与《国际建协认同书》配套的《政策推荐导则》，这些《政策推荐导则》对《UIA认同书》中的各条内容作了更具体的规定。

《国际建筑师协会认同书》及其相应的《政策推荐导则》旨在保证全球建筑师的职业资格质量和服务水平，同时为各国各地区之间建筑师资格的互认谈判及建筑师专业服务的全球化提供了一个基本框架和共识。

在国际上，除了各国立法机构和行政当局制定的有关法规和标准外，一些国际专业团体也编制了有关规定，其中有些规定被视为国际惯例和"国际游戏规则"而普通采用，例如国际咨询工程师联合会(FIDIC)制定的《土木工程施工合同条件》在国际工程建设中被广泛采用。国际建筑师协会所制定的《UIA认同书》及一系列《政策推荐导则》在建筑师服务贸易走向全球化的进程中正越来越显示它的指导意义和应用价值。到目前为止世界上多个国家正根据《UIA认同书》达成了互认协议，世界上两个建筑师最多的国家和地区：美国全国注册建筑师委员会(NCAQB)和美国建筑师学会(AIA)与欧洲建筑师委员会(ACE)于2002年12月已签订了"合作认同书"。

中国作为一个发展中国家，建立职业制度还刚刚起步，有关职业的标准和规定还很不完善，很多领域几乎是空白，从这个意义上说，国际建筑师协会关于建筑师推荐国际标准对于我国建筑师职业标准的建立和完善有重要的参照意义。

目前国外和境外建筑师在中国的建筑实践相当广泛，许多国内的设计单位特别是大中型设计院，都在与国外和境外建筑师合作设计，对于这些设计单位和从事合作设计的建筑师，《国际建筑师协会认同书》和《政策推荐导则》是双方都可以认同的"国际语言"，它对于协调合作双方的各类问题都有帮助和实际的运用价值。

中国建筑师走向世界的还不多，但随着我国经济发展加快和建筑师水平的提高，在未来的几十年中，将会有越来越多的中国建筑师会走出国门，提供跨境建筑师专业服务，从这个意义上说，《国际建筑师协会的认同书》等文件对我们有更长远的意义。

需要说明的是，和所有的标准一样，国际建筑师协会所制定的推荐国际标准和导则并不是一成不变的。随着时间的推移和实践中出现的新问题，这些标准和导则只要得到国际建筑师协会理事会和代表大会的批准就可以进行修订。本书中收集的是到目前为止所制定的文件。有些《政策推荐导则》尚在制定过程中。

为了更好理解和对照推荐国际标准，本书中附上了英文文件，原文件有英文和法文两种版本，并已经译成多国文字，可查阅网站：www.aia.org/institute/UIA

在本书附录中收录了两篇文章和一篇讲话稿及一份提案。第一篇是席勒先生所写，第二篇是张钦楠先生所写，第三篇是2002年7月在柏林举行的世界建筑师大会上由WTO的国内法规工作组秘书D·B霍纳克先生的讲话稿，第四篇是世贸组织多哈回合谈判（将于2005年1月结束）澳大利亚政府关于建筑师服务贸易谈判的提案，这四篇的内容会有助于对国际建筑师协会制定的推荐国际标准的背景及建筑师服务贸易的全球化进程有更多的了解。

最后，对于促成UIA职业标准形成的UIA前主席斯哥塔斯先生，副主席哈克尔先生及理奎特秘书长致谢，对于席勒和张钦楠先生及为制定这些标准作出贡献的许多国家建筑师同行致谢。对于促成本书出版的中国建筑工业出版社和王莉慧编辑致谢。

许安之
2003年12月于深圳

目 录

前言

国际建筑师协会(UIA)关于建筑实践中职业主义的
推荐国际标准认同书…………………………………………………9

国际建筑师协会关于建筑实践中职业主义的推荐国际标准

 关于建筑教育评估的政策推荐导则……………………………27
 关于建筑实践经验、培训和实习的政策推荐导则……………32
 关于注册、执照、证书的政策推荐导则………………………37
 关于道德和行为标准的政策推荐导则…………………………45
 关于继续职业发展的政策推荐导则……………………………51
 关于在东道主国家建筑实践的政策推荐导则…………………54
 关于知识产权和版权的政策推荐导则…………………………59

附录：

 1. 建筑与职业制度/詹姆斯·席勒　章岩译……………………67
 2. 全球化时代的职业精神/张钦楠…………………………………82
 3. 较自由的服务贸易市场/D.B霍纳克……………………………87
 4. 多哈回合谈判澳大利亚政府关于建筑服务贸易的提案………93

UIA Accord on Recommended International Standards of
Professionalism in Architectural Practice……………………………95

Recommended Guidelines for the UIA Accord On Recommended
International Standards of Professionalism in Architectural
Practice…………………………………………………………………117

 1. Policy on Accreditation/Validation/Recognition………………117
 2. Policy on Practical Experience/Training/Internship……………124
 3. Policy on Registration/Licensing/Certification of the Practice of
  Architecture…………………………………………………………130
 4. Policy on Ethics and Conduct……………………………………140
 5. Policy on Continuing Professional Development………………148
 6. Policy on Practice in a Host Nation………………………………151
 7. Policy on Intellectual Property and Copyright…………………157

国际建筑师协会（UIA）

关于建筑实践中职业主义的
推荐国际标准认同书

第三版

1999年6月28日在中国北京举行的UIA第21届代表大会通过
前言于2002年7月27日在德国柏林举行的UIA第22届代表大会通过

UIA 职业实践委员会联合书记处

联合主任
美国建筑师学会
詹姆士·A·席勒，FAIA
1735 New York Avenue, NW Washington,
DC 20006
电　话：202 - 6267315
传　真：202 - 6267421

联合主任
中国建筑学会
张钦楠，副理事长
中国北京西城区百万庄
邮　编：100835
电　话：8610 - 88082239
传　真：8610 - 88082222

序　言

　　作为专业人士的建筑师，有责任关注他们所服务的社会，这种责任要比建筑师个人的兴趣和利益以及他们的甲方利益更为重要。

　　当今世界专业人士的服务贸易在迅速增加，建筑师经常不仅仅服务于他们自己的国家和地区，国际建筑师协会相信，需要一套国际的建筑师实践的职业标准。符合 UIA 认同书中规定要求的建筑师，包括其所受的建筑学教育，具有的能力和职业道德，能保护他们所服务的社会的最大利益。

引 言

国际建筑师协会理事会于1994年设立了职业实践委员会,并批准了它的工作计划。在1993—1996年的这3年的时间内,UIA的职业实践委员会经过25个月紧张的工作后,于1996年7月在西班牙的巴塞罗那由UIA理事会推荐,UIA代表大会一致采用了该委员会汇编的第一版《UIA关于建筑实践中职业主义的推荐国际标准认同书》。由此,UIA的认同书被确立为UIA及UIA职业实践委员会工作的指导性方针。

第一版的认同书发给UIA所有成员组织,并要求它们合作和参与本文件的继续完善,以便提交给UIA1999年于北京召开的第21届代表大会。1997—1999年的计划主要集中在对于UIA理事会成员、UIA成员组织和职业实践委员会成员对于认同书及其方针的意见和建议。根据这些评论和委员会讨论的结果,第一版认同书被修改,并进一步制定若干政策的导则作为对认同书政策框架的补充。

认同书和导则尊重UIA各成员国的主权,允许对等原则留有灵活性,并应当允许某一UIA成员组织根据本国具体条件制定出附加要求。

认同书是非强制性的,可由成员组织在自愿的基础上使用。认同书是由国际建筑师大家庭共同合作努力的结果,以建立客观的标准和实践,更好地为建筑领域服务。认同书和导则的目的是规范什么是建筑学专业的最佳实践以及什么是建筑师职业所追求的标准。这些文件是动态生长的,将会不断接受检验,当成员组织的意见和经验占主导地位时,条文可被修订。在尊重UIA成员国自主权的同时,鼓励他们采纳认同书和导则,如果有可能,探寻对各自国家现有惯例和法律条文的修改。

认同书和导则是为政府间、有关谈判各方和其他为进行互认建筑服务谈判的各方提供可操作的导则。认同书和导则将会使谈判各方更容易达成互认协议。GATS 第七款认同的，相互承认最普遍的方式是通过双边协议。由于在教育和考试标准、经验要求和法规影响等方面的差别，使多边基础的相互承认极难实现。双边谈判则有利于集中解决两个特定环境中存在的关键问题。一旦成功，双边协议可以导致在更广泛的范围内实现相互承认。

本认同书开始有一"职业精神的原则"条文；继之以一系列政策条文。每项政策均包括：政策定义、背景说明和政策本身三个部分。

1999 年 7 月于北京召开的第 21 届代表大会一致通过认同书。大会决议的副本见附录 A。

引言
序言
目录
国际建筑师协会关于建筑实践中职业主义的推荐国际标准认同书

 职业精神原则   1
 政策   2
  ◆ 建筑学实践   2
  ◆ 建筑师   2
  ◆ 对一名建筑师的基本要求   3
  ◆ 教育   4
  ◆ 建筑教育的评估和认证   5
  ◆ 实践经验/培训/实习   5
  ◆ 职业知识和能力的证明   5
  ◆ 注册/执照/证书   6
  ◆ 取得委托   6
  ◆ 道德与行为   7
  ◆ 继续职业发展   7
  ◆ 实践范围   8
  ◆ 实践形式   8
  ◆ 在东道国的实践   9
  ◆ 知识产权和版权   9
  ◆ 职业团体的作用   10

附录 A   关于正式通过 UIA 建筑实践中职业主义推荐
国际标准认同书的决议（第 17 号）   11
 注：认同书中下列政策已编成了导则：
 建筑教育评估   13
 实践经验/培训/实习   18
 职业知识和能力的证明*
 注册/执照/证书
 按质选择建筑师（QBS）*
 道德和行为标准

职业继续教育	37
在东道主国家的实践	
知识产权和版权	40

带 * 的三个导则因为在 UIA/AIA 网上未公布而未被收录在本书内（编译者注）

国际建筑师协会
关于建筑实践中职业主义的推荐国际标准认同书

职业精神原则

建筑师职业的成员应当恪守职业精神、品质和能力的标准,向社会贡献自己的为改善建筑环境以及社会福利与文化所不可缺少的专门和独特的技能。职业精神的原则可由法律规定,也可规定于职业行为的道德规范和规程中。

专长:

建筑师通过教育、培训和经验取得系统的知识、才能和理论。建筑教育、培训和考试的过程向公众保证了当一名建筑师被聘用于完成职业服务时,该建筑师已符合为适当完成该项服务的合格标准。此外,多数建筑师的职业协会以及 UIA 均被赋予维持和提高其成员对建筑艺术和科学的知识、尊重建筑学的集体成就,并为其发展作出贡献的责任。

自主:

建筑师向业主及/或使用者提供专业服务(原词也可为"咨询",但和我国的理解不同),不受任何私利的支配(尤指拿施工单位、材料设备供应商等"好处费")。建筑师的责任是,坚持以知识为基础的专业判断分析,在追求建筑的艺术和科学方面,应当优先于其他任何动机。

建筑师还要遵守从精神到有关建筑师事务的法律条文,并周到地考虑到其职业活动所产生的社会和环境影响。

奉献:

建筑师在代表业主和社会所进行的工作中应当具有高度的无私奉献精神。本职业的成员有责任以能干和职业方式为其业主服务,并代表他们作出公平和无偏见的判断。

负责:

建筑师应意识到自己的职责是向业主提出独立的(若有必要时,甚至是批评性的)建议,并且应意识到其工作对社会和环境所产生的影响。建筑师和他们所聘用的咨询师只承接他们在专业技术领域中受过教育、培训和有经验的职业服务工作。

UIA 通过其成员国组织及其职业实践委员会的计划，寻求确立保障公众健康、安全、福利和文明所需要的职业精神原则以及职业标准，并支持相互承认职业精神和能力的标准是符合公众利益并维护其职业信誉的立场。

UIA 的原则与标准旨在对建筑师进行完善的教育和实践培训，使其能达到基本的职业要求。这些标准承认不同国家的教育传统，因而也允许对等性因素。

政　　策

建筑学实践

定义：

建筑学实践包括提供城镇规划以及一栋或一群建筑的设计、建造、扩建、保护、重建或改建等方面的服务。这些专业性服务包括(但不限于)规划，土地使用规划，城市设计，提供前期研究、设计、模型、图纸、说明书及技术文件，对其他专业(咨询顾问工程师、城市规划师、景观建筑师和其他专业咨询顾问师等) 编制的技术文件作应有的恰当协调以及提供建筑经济、合同管理、施工监督与项目管理等服务。

背景：

建筑师从古以来就从事其艺术和科学的实践。我们今天认识到的建筑行业经历了巨大的发展和变化。对建筑师作品的形象要求变得更为苛刻，业主要求和技术进步变得更为复杂，社会和生态的使命变得更为迫切。这些变化产生了服务内容的变革以及在设计和施工过程中与更多方面的合作。

政策：

在制定 UIA 国际标准时采用以上"建筑学实践"的定义。

建筑师

定义：

"建筑师"的定义，通常是依照法律或习惯专门给予一名职业上和学历上合格并在其从事建筑实践的辖区内取得了注册/执照/证书的人，在这个辖区内，该建筑师从事职业实践，采用空间形式及历史文脉的手段，负责任地提倡人居社会的公平和可持续发展，福利和文化表现。

背景：

建筑师属于在较大范围内国营和私营部门的财产开发、建筑和施工经济行业的一部分，他们从事委托、保护、设计、建造、装修、理财、管理和调节我们的建造环境以满足社会需要。建筑师在多种状况和组织形式下工作。比如，可以自营或受雇于私营或国营部门。

政策：

在制定 UIA 国际标准时采用以上"建筑师"的定义。

对一名建筑师的基本要求

定义：

按以上定义的建筑师，其取得注册/执照/证书的基本要求是必须掌握下列各种需要通过教育和培训而取得的可证明的知识、技能、能力和经验，有从事建筑实践的职业资格。

背景：

1985 年 8 月，有一批国家首次共同拟定了一名建筑师所应具有的基本知识和技能，包括：

• 能够创作可满足美学和技术要求的建筑设计；

• 有足够的关于建筑学历史和理论以及相关的艺术、技术和人文科学方面的知识；

• 与建筑设计质量有关的美术知识；

• 有足够的城市设计与规划的知识和有关规划过程的技能；

• 理解人与建筑、建筑与环境、以及建筑之间和建筑空间与人的需求和尺度的关系；

• 对实现可持续发展环境的手段具有足够的知识；

• 理解建筑师职业和建筑师的社会作用，特别是在编制任务书时能考虑社会因素的作用；

• 理解调查方法和为一项设计项目编制任务书的方法；

- 理解结构设计、构造和与建筑物设计相关的工程问题；
- 对建筑的物理问题和技术以及建筑功能有足够知识，可以为人们提供舒适的室内条件；有必要的设计能力，可以在造价因素和建筑规程的约束下满足建筑用户的要求；必须要有在造价和建筑法规约束下满足使用者要求的设计能力；
- 对在将设计构思转换为实际建筑物，将规划纳入总体规划过程中所涉及到的工业、组织、法规和程序方面要有足够的知识；
- 有足够的对项目资金、项目管理及成本控制方面的知识。

政策：

UIA 在制定国际标准时，将上述的基本要求作为基础，并寻求保证这些专门的要求在教育课程中得到强调，UIA 还寻求保证这些基本要求得到经常的审查，使其在建筑师职业和社会的演变中能保持适应。

(*根据欧洲共同体委员会 EEC 指令 85/384)

教育

定义：

建筑教育应保证所有毕业生有能力进行建筑设计，包括其技术系统及要求，考虑健康、安全和生态平衡，理解建筑学的文化、知识、历史、社会、经济和环境文脉，理解建筑师的社会作用和责任，并具有分析和创造的思维能力。

背景：

在多数国家，建筑教育通常由一所大学以 4~6 年全日制的学术性教育提供（在有的国家，在成功地完成学业后，还要有一段实践经验/培训/实习），但由于历史原因也有重要的差别（非全日制，工作经历等）。

政策：

按照 UIA/UNESCO 的《建筑教育宪章》，UIA 提倡建筑师的教育（不包括实践经验/培训/实习）应不少于 5 年，主要以全日制为基础，大学应通过评估，建筑学教育计划须通过评估，允许在教学思想和对地方文脉的反映上有差别，并考虑灵活对等的原则。

建筑教育的评估和认证

定义：

评估指的是确认某一教育计划符合已确定的成绩标准的过程，其目的是保证维持与改善适宜的教育基础。

背景：

由一个独立的机构确立评估的合格准则及程序有利于发展完整协调的建筑教育计划。经验证明，通过定期的、外来的监督以及在某些国家中加上内在的质量保证检查可使教育标准取得协调与改进。

政策：

建筑学课程必须由本大学以外的一个独立机构在通常不超过5年期内进行评估与认证。UIA与有关国家高等教育机构一起，按照程序编制知识连贯、面向实践的建筑师学历性职业教育标准。

实践经验/培训/实习

定义：

实践经验指的是在取得职业学位之后，在为取得注册的考试之前的有指向性和结构良好的建筑职业活动，是指在建筑学教育过程中或者是在获得建筑学专业学位后，但在取得注册建筑师职业资格前的有指导性、组织性的建筑学实践活动。

背景：

为保护公众，申请取得注册/执照/证书者应当受到正式教育并通过实践培训对其学历教育进行补充和充实。

政策：

建筑学毕业生在取得注册/执照/证书之前，应完成2年的合格的经验/培训/实习（今后目标是3年），才能作为建筑师执业，同时允许有对等的灵活性。

职业知识和能力的证明

定义：

每名申请取得建筑师注册/执照/证书者，应向有关当局证明自己具有合格的职业知识和能力水平。

背景：

一名建筑师知识和能力的保证只有在他（或她）经过所要求的教育和培训（实践经验/实习）并证明他（她）具有最低限度从事全面建筑实践的知识能力后才落实。这些要求要通过考试及/或其他来证明。

政策：

要有足够的证据证明一名建筑师所具有的知识和能力。这种证明应当包括在实践经验/培训/实习后至少通过一次考试。在考试范围以外的建筑师职业实践的能力中某些必要的部分（包括诸如业务管理和相关法律要求）要通过其他足够的证据证明。

注册／执照／证书

定义：

注册/执照/证书是官方对一名个人具有作为建筑师实践资格的法律承认。法律同时禁止不合格者履行此种职能。

背景：

考虑到公众在高质量的、持续发展的建造环境中的利益，以及与该环境相关的危险和后果，建筑师的服务只能由合格的人员担任，以便向公众提供足够的保障。

政策：

UIA促进在世界各国建立建筑师的注册/执照/证书制度。这种制度的条文应当由法律规定。

取得委托

定义：

建筑服务取得委托的过程。

背景：

建筑师（通过其行为准则）应当在自己利益之前首先坚持维护业主和社会的利益。为了保证按公众利益所要求的标准，完成建筑任务需要足够资源，他/她们通常按照规定的或推荐的职业收费标准取得报酬。

尽管有一些国际法则，如国际贸易组织的一般委托协议和欧洲共同体的服务指南等，旨在保证对建筑师进行客观和公正的选择。然而，在公共和私人工程中，单纯地以设计费高低而选择建筑师的倾向越来越严重。这种以设计费为基础的做法将迫使建筑师减少自己对业主的服务，这反过来会影响建造环境的质量、福利设施和社会及经济价值。政策为保证建造环境的生态持续发展并保障社会、文化和经济价值，政府应当使指定建筑师的过程能导向选择最适宜本项目的建筑师，在有足够资源的条件上，经过协商，以下方法可最好地实现此目的：

- 根据 UNESCO – UIA 确定的有关国际竞赛导则，并由国家主管部门及/或建筑师职业协会所批准的建筑设计竞赛。
- 以一份完整地界定建筑师服务内容的任务书为基础的直接谈判。
- 按国际建协有关导则进行的基于质量的选择程序（QBS）。

道德与行为

定义：

一项道德与行为规范确立了建筑师在实践行为中指导其行为的职业标准。建筑师应当遵守在其执业的辖区内的道德和行为规范。

背景：

道德和行为规范的主要目的是保护公众，关心较弱小的阶层以及一般的社会福利，推进建筑师的职业利益。

政策：

现有的 UIA 关于咨询服务的国际道德规范继续有效。鼓励 UIA 成员组织在自己的道德和行为规范中引入推荐标准的认同书和实施一项其成员应遵守在其提供职业服务的国家和辖区现行的道德和行为规范的要求，只要该规范的规定不受国际法或建筑师本国法律的禁止。

继续职业发展

定义：

继续职业发展（CPD）是一个终身的学习过程。它能维持、改善和增加建筑师的知识和技能，以保证其持久的职业能力。

背景：

越来越多的职业团体和管理部门要求其成员付出时间（典型的是每年35小时）学习，以维持其现有的技能，扩大其知识并探索新概念。现在日益需要跟上新技术、新的执业方法和新的社会和生态条件。继续职业发展可以由职业团体规定作为更新或继续会籍的条件。

政策：

UIA鼓励其成员组织从公众利益出发建立继续职业发展制度，并作为会员的责任。建筑师应当保证他们能够提供合格的服务和行为规范，应保持"建筑师的基本要求"中所描述的各种知识标准及其未来转变所要求的相关标准。同时，UIA应关注以继续职业发展为更新注册的条件，并推荐有关导则以促进国家间的对等承认，继续完善此项政策。

实践范围

定义：

提供有关土地使用规划、城市设计和建筑项目的设计与管理服务。

背景：

随着社会的演变，城市与建造环境变得越来越复杂。建筑师要涉及越来越多的关于城市、美学、技术和法律方面的非常广阔的问题。现已证明，需要一种对建筑设计的综合途径，以保证法律、技术和实际的要求得到解决，社会的需求得到满足。

政策：

UIA鼓励并促进建筑实践的范围不断扩大，它只受行为规范的限制。努力保证这种范围的扩大，相应必要的知识和技能也要拓宽。

实践形式

定义：

建筑师提供职业服务的合法实体。

背景：

传统上，建筑师以个人、合伙人或受雇于公私机构的方式从事实践。近期，实践的要求导向多样的结合，例如，有限和无限责任公司、合作经营、以大学为基地的设计所、社区建筑师等，但并不是所有各国都允许这些形式。有些结合还包括其他专业。

政策：

建筑师应当可以在提供服务的国家内以任何法律允许的形式实践，但要受道德行为要求的约束。UIA认为有需要时，可以发展或修订其政策与标准，以考虑其他实践形式和多种地方条件，只要这些形式可在保证公众利益的前提下，发挥建筑师职业的积极性和创造性的作用。

在东道国的实践

定义：

在东道国的实践指的是一名个体建筑师或建筑师的企业实体在另一个国家寻求任务委托或被委托设计某项目的情况。

背景：

对建筑师责任的流动性及其在外国管辖地区内提供服务的兴趣正在日益增加。同时也出现对本地环境、社会和文化因素及道德和法律准则的关注。

政策：

建筑师在他们未曾注册的国家提供一个项目的建筑服务时，建筑师应当与当地建筑师合作以保证恰当地和有效地理解当地的法律、环境、社会、文化和传统因素。合作的条件应当由双方依据UIA的道德准则和地方法律法规确定。

知识产权和版权

定义：

知识产权包括了专利、版权和商标三个法律领域。它涉及（有的国家用法律来保证）设计师、发明家、作家和制造者对其概念、设计、发明、创作的权利以及对产品及服务的来源的识别。

背景：

尽管许多国家对建筑师的设计有法律保护，但这种保护往往是不够的。有时建筑师与一位可能的业主讨论某些观点和构思，之后不但没有被委托，而且还发现其观点已被毫无报酬地被别人使用，这种现象并不少见。建筑师的知识财产，在某种程度上，得到国际法规的保护。GATS有一项涉及到与贸易有关的知识产权的协议，包括对仿冒产品的内容(TRIPS)。1955年9月16日的世界版权大会具有国际意义。在欧洲，1886年的伯恩修订协议对其多数国家继续有效。

政策：

UIA成员国组织的国家法律应确保一名建筑师在不损害其权利和责任的条件下从事职业实践，并有权保持他/她的作品的知识产权/版权。

职业团体的作用

定义：

职业通常由一制定标准（如教育、道德规则、职业标准等）的管理机构所控制。这些规则和标准是为公众利益而不是为成员的私人利益而制定的。在某些国家，由法律向某些职业保留某类工作，不是为了该职业的私利，而是为了保护公众。它们只能由已取得必要的培训、标准和纪律的人来承担。为推进建筑学的发展，提高知识水平并保证成员按一定的标准完成任务，保护公众利益，成立了职业团体。

背景：

根据一个国家对建筑师称号或职能（或二者具备）的保护，职业团体的作用和责任也有较大的不同。在有些国家，管理单位也代表职业，而在其他国家，这些功能是分开的。通常职业团体的成员要维持一定的标准，这是由成员遵守职业团体所制定的行为规范和完成有关部门要求，如继续职业发展等而实现的。

政策：

在没有职业团体的国家中，UIA鼓励建筑职业的成员为公众的利益而成立此类团体。职业团体应寻求保证其成员遵守UIA国际标准，达到UIA/UNESCO的建筑教育宪章和UIA道德和行为标准，按《基本要求》（现有的和今后发展）不断地增进其知识和技能，并对建筑文化、建筑知识及其服务的社会的发展作出贡献。

附录 A

关于正式通过 UIA 建筑实践中职业主义推荐国际标准认同书的决议

（第 17 号）

国际建筑师协会第 21 届代表大会
中国　北京　1999 年 7 月 28 日

代表大会一致通过 UIA 关于建筑实际中职业主义推荐国际标准认同书(第二版)，将它作为各成员组织在制定和重审它们自己标准时的参照文件。认同书和导则将会使 UIA 成员组织更容易谈判互认协议。

代表大会要求将认同书传达到国际建筑师协会的所有成员组织，要求它们参与和合作，在 2002 年 UIA 第 22 次代表大会上进一步发展这个政策框架。

大会委托 UIA 理事会批准认同书中的导则文件并将它们推荐给 UIA 各成员组织。

大会认为在不同成员组织间存在着不同的文化、不同的实践、不同的条件，因此应当鼓励在使用这些文件作为参照文件时要适应当地条件。

UIA 代表大会深知在互认协议谈判中每一个 UIA 成员组织的主权都应当被尊重，应注意这些导则在等效和互惠的原则上允许有灵活性，允许反映 UIA 的一个成员当地的附加要求。

大会授权 UIA 主席和秘书长将 UIA 认同书提交给世界贸易组织和其他有关的组织和机构作为互认谈判的基础，并授权给他们，当某国的 UIA 成员组织有专门要求时，将 UIA 认同书提交给该国政府。

代表大会要求职业实践委员会分析在大会期间所反映的所有意见以及在布拉格召开的职业实践委员会（1999 年 10 月）上所反映的意见，检查是否应当融合在北京所通过的文件中。

大会授权 UIA 理事会制订一项政策条文，将 UIA 认同书和导则文件与各个方面作沟通和介绍。

大会向 UIA 各成员组织建议，在使用这些标准之后，将它们的体会和经验反映给职业实践委员会，使这些基本文件被考虑以后得到改进和修订。

国际建筑师协会
建筑实践中职业主义的推荐国际标准

关于建筑教育评估的政策推荐导则

1998 年 3 月制订
1998 年 4 月 23 日修订
1998 年 12 月 10—12 日再修订
1999 年 6 月通过

~~~~~~~~~~~~~~~~~~~~~~~~~~~~~~~~~~~~~

国际建筑师协会职业实践委员会秘书处

美国建筑师学会
联合主任　J.A.席勒
1735 New York Avenue, NW
Washington, DC 20006
电　话：202 - 6267315
传　真：202 - 6267421

中国建筑学会副理事长
联合主任　张钦楠
中国北京西城区百万庄
邮　编：100835
电　话：8610 - 88082239
传　真：8610 - 88082222

认同书中关于建筑教育评估和认证的政策

建筑学课程必须在合理的时间段内（通常不超过 5 年），由大学以外一个独立机构来评估和认证。UIA 与有关国家的高等教育组织一起编制在知识上连贯并面向实践的建筑师学历性职业教育标准。

引　言

建筑学教学大纲的评估，不管是教育机构评估还是完全由相关机构评估，其目的主要是保证公众利益。教学大纲指导下的合格毕业生能胜任建筑实践中对于设计、技术、职业技能和美学的要求。

任何评估政策的原则是：允许评估方式的灵活性；评估机构的独立性；不断追求教育和评估过程自身的高标准。评估要满足上述规定并符合教学大纲要求，评估专家由校外聘任，他们应当有经验并受过训练，能胜任评估建筑学教育计划并能提出方向和建议。校外评估专家可以由管理建筑教育的政府机构聘任，也可以由独立的建筑职业团体聘任，还可以由被评估的建筑系提名，或其他合适的方式。评估专家的聘任可以根据公立和私立大学而不同。所指的独立机构可以是一个职业团体，如建筑师学会或其他非政府的建筑师组织或建筑系组织，它可以是全国性的，也可以是省一级的，还可以是校外专门的评估组织。评估过程中如期寄送一套满足要求的评估方法，包括对建筑系近一年所有学生作业的检查。应承认公立和私立大学的差别。必须保持评估过程和最后结果的连续性。

评估程序分下列情形：建筑系已评估过；未评估过；教学计划有变动。每种情形下，评估专家在访系前先得到自评估文件，到系后检查论文和材料，设计课教学计划，其他课程教学计划，课程作业实例。评估专家与学生和教师会面，他们还可能检查教师的教学方法、专业和研究成果。评估专家将向建筑系提供关于评估结论的报告，推荐评估通过并提出教学大纲的改进建议，或针对评估提出强制性条件。

1. 对课程和考试的评估标准

关于合格建筑师的核心知识和技能要求，有关高等教育组织和 UIA 认同书对建筑师的基本要求已作了规定。

建筑师通过教育、训练、经验和建筑专业学习教育获得的技能可帮助建筑学专业的学生获得能力、知识、理解和技能。这些要求应在教学计划中体现。

UIA 主张建筑师的教育不应少于 5 年，主要以全日制为基础，大学和建筑学专业均需通过评估。允许在等效性上有弹性。在某些国家，建筑教育之后有一段实际工作和实习的时间。在这种教育和训练过程中，建筑系学生所具有的 UIA 基本要求所列出的能力水平要好，教育评估将会考虑合适的学习时间和不同的水平。

对建筑师的知识和能力的要求反映了社会需求的变化，UIA 会及时检查 UIA 认同书的各推荐导则，使它与时俱进。

对于在不同国家和在不同时间，评估标准对于技能的侧重程度也会变化。在不同国家，由于传统和取向不同，建筑教育机构可能对各种评估标准有它们自己的侧重面，通常会受该国建筑师具体的工作任务的影响。在每种情况下，教育计划须与建筑师基本要求及其派生出来的要求相结合。评估标准包含对教育计划的检查，教育计划随学习阶段而变化，如期中、期末、实习前后等。

2. 评估方法

评估需由被评估单位外的独立机构来做。评估机构必须训练有素，有经验，能够胜任。担任评估工作的个人应有建筑设计工作的实践，道德品质好，有训练经验。常常评估组成员由一个以上类别的组织机构推荐。在任一情况下，建筑师职业团体的成员须参加评估，这将有助于以更客观更综合的建筑学眼光来评估。

当有教育机构参与评估程序时，被评估的教育机构不能参加自己的建筑学专业教育评估。

3．评估程序

评估内容和细则由一个评估委员会制定，并根据该国的文化和教育实践而变化。评估还应根据被评估的建筑系的情况而不同。有的系是在教育计划开设之前，有的是第一次评估，有的是已设立了一段时间并已经评估过，有的是已评估过但未通过，或者以前的评估已撤消，为了以后而重新评估。

评估程序还会因评估过程为一阶段或多阶段而不同。在某些国家，评估程序分三阶段，即3年和5年课堂教育后的评估以及实践训练结束后的评估。在其他国家，评估过程在一或两个阶段内。

评估程序包含评估专家们对教育计划内容及该计划中已达到标准的评估。检查教育计划，课程设置和详细内容和考卷，校外督察员的报告，该系的自我评估。评估访问期间与系主任、教师和学生会面，检查学生作业和教学设施。

当一个系对已有课程实施重大改变或准备设立新课程时，最好由一个独立机构作一个初步评审，对新课程的内容、结构、资源等作评审，以便使该课程达到评估要求。建筑系应将准备新设的课程内容、原理和方法、教学计划等信息提供给评审专家，介绍的内容包括课程框架和详细内容、课程要求、讲课要点和每一部分的学时分配。

4．评估材料和评估访问方式

无论是第一次评估还是续评估，被评估的建筑系应准备的材料是：

- 上级单位（大学）的简要介绍，所在学校是全国性的、地区性的、还是城市性大学。
- 简述课程简要历史。
- 建筑教育的指导思想和培养目标。
- 影响课程的学生基础背景特点。

- 教师简历，含非教学的活动和其他工作任务，如研究、著作、职业实践、社会工作。
- 有形资源的介绍：设计教室、教学空间、教学设备、实验室、工作室、工场、图书馆设施、资源中心、计算机和信息系统。
- 管理和决策框架简介。
- 学历教育计划的完整介绍，包括教学计划框架和要求，以及对毕业生的其他要求，讲座计划，设计课的详细介绍，课堂笔记的复印件。
- 注册学生的统计信息，毕业生数量，教师数量，师生比率。
- 本系教育政策的自我评估。应考虑以前的评估报告，以后的发展。自评报告还应包含校外督察员的报告、资源的变化、主要课程目标的评估、课程特色，以及其他有关事项。

评估组访问时在现场检查教育计划，被评估的建筑系有一个至少12个月内的学生作业展览，这对评估会很有帮助。该展览内容应包括每一年的设计课程作业，以便尽可能展现这整个教育计划的发展过程。每一年所写的和所画的作业都应展览，使得学生的每一方面的水平是否符合建筑师要求都可进行评估。展出的作业应包括各门课程的最高成绩，平均成绩以及刚刚够及格线的成绩，这些作业还应有各课程各年级的考试和考查成绩记录来补充。

当现场评估视察时，评估组可以与教师包括学院院长、系主任和校外督察员举行会面讨论。评估组还可与学生座谈，评估组一方可以是评估组全体或者是评估组个别成员，讨论内容可包括教学方法的评议、设计题目的内容、讲课课程及其专门的教师，未来的发展等。

5．评估报告的程序

评估组应提交一份对建筑学专业评估访问的结论性报告。该报告对被评估建筑系的文字材料进行评估和补充，对检查各课程及其学生的表现所体现的教育水平提出评估意见。程序可以包括如何保证此报告符合事实，保持秘密，并供各有关部门阅存。评估报告通常推荐评估通过后的有效期不超过5年。

国际建筑师协会
建筑实践中职业主义的推荐国际标准

关于建筑实践经验、培训和实习的政策推荐导则

1998 年 4 月初稿
1998 年 12 月 10—12 日修订
1999 年 6 月通过

国际建筑师协会职业实践委员会秘书处

美国建筑师学会
联合主任　J.A. 席勒
1735 New York Avenue, NW
Washington, DC 20006
电　话：202 - 6267315
传　真：202 - 6267421

中国建筑学会副理事长
联合主任　张钦楠
中国北京西城区百万庄
邮　编：100835
电　话：8610 - 88082239
传　真：8610 - 88082222

认同书中关于建筑实践经验、培训和实习部分的政策

建筑学毕业生在为取得注册/执照/证书的考试之前，应至少完成2年的合格的经验/培训/实习（今后目标是3年）才能作为建筑师执业，同时允许有对等的灵活性。

导　则

1. 实践经验/培训/实习的时期

按以下要求在政策规定的期限内取得的经验应当在注册/执照/证书的考试之前完成。至少有一半是在完成经过评估的正式教育之后取得的。

2. 实践经验/培训/实习期的目标

建筑实践经验/培训/实习（以下简称实习）期的目标是：
- 向实习人提供机会取得建筑实践中基本的知识和技能。
- 保证实习人的实践活动和经验用一种标准的方式予以记录。
- 帮助实习人取得建筑实践中广泛的经验。

3. 经验类型

实习人应在一名建筑师的指导下，在以下四个类型的每一项中至少一半的经验范围中取得实践经验和培训：

3.1　项目和办公管理
- 与业主会晤
- 与业主讨论设计任务书及初步设计图纸
- 将业主要求定型
- 合同前的项目管理
- 确定合同条件

- 起草通信文件
- 协调咨询师的工作
- 办公及项目会计系统

3.2 设计与设计文件的编制
- 场地调查和评价
- 与有关部门会晤
- 对与相关规程关联的评价
- 制作方案及发展设计图
- 根据法规要求检查设计
- 制订预算、成本计划和可行性研究

3.3 施工图文件
- 制作施工图和施工说明书
- 根据进度表及成本计划监督设计进度
- 根据法规要求检查设计
- 协调咨询师的设计文件
- 协调合同图纸及说明书

3.4 合同管理
- 现场会议
- 施工检查
- 向承包人发出指示、通知和证明
- 向业主报告
- 变更和付款管理

4. 培训记录

实习人应保持一书面记录，以标准格式或记录本记载在实习期间接受的培训、经验和补充教育。

本记录应按上述第二款中提出的实习范围排列。记录中应描述所经历的活动的性质和活动时间，其中每项均应由指导的建筑师签字，作为实习人取得经验的真实记录。

这一标准格式或记录本应在注册/执照管理部门提交审查，作为已经进行或已经完成所要求的实践经验/培训/实习的证据。

5. 监督人

实习人应在监督下取得经验。监督人应是本国或本州的注册或执照建筑师，可以是实习人的雇主或由其定期报告者。

6. 核心能力要求

在实践经验/培训/实习期终了时，实习人应当显示或能够显示具有以下知识和能力。

6.1 建筑实践
- 对国内外建筑职业总的了解
- 对道德标准的知识和重视
- 对当地建筑职业的认识
- 对当地建筑业和建筑法的认识
- 指导和协调咨询师
- 办公管理和体制
- 法律方面的实践
- 职业责任、风险管理和保险

6.2 项目管理
- 确定和建立与业主的协议
- 项目活动和任务的进度表
- 评价规范、规程和立法
- 项目资金和成本控制
- 项目取得和合同制度
- 纠纷的解决
- 分包合同的管理
- 项目管理和监督系统

6.3 设计前期工作与场地分析
- 确立、分析和记录与项目有关的环境问题
- 确立和清晰定义设计任务书
- 确立、分析和记录场地条件

6.4 建筑服务及其系统
- 在项目设计和文件编制中，协调建筑服务及其系统的设计和

文件编制

6.5 方案设计
- 分析业主的任务书，通过假设、评价和再评价的过程，产生可用的设计解决方案
- 用图来表现设计方案
- 提出初步设计方案，并与业主和其他有关部门达成一致

6.6 深化设计和编制设计文件
- 调查和确立特定的空间、组织和室内外流通要求
- 考虑和决定结构和建筑服务系统、材料和构件的处置
- 制作图纸和文件以充分描述发展的设计提案供业主和其他有关部门批准

6.7 施工文件
- 为一建筑项目研究、分析和选择适宜的材料和系统
- 编制正确、一致和完整的施工图纸、施工说明书、进度表，描述建筑构件、部件、装修、配件和系统的大小和位置

6.8 合同管理
- 编制招标文件
- 评标并提出推荐
- 确认项目合同
- 管理项目合同
- 监督合同条件和遵守有关部门的要求
- 检查和评价建筑工程以确保其符合合同文件的要求

国际建筑师协会
建筑实践中职业主义的推荐国际标准

关于注册、执照、证书的政策推荐导则

1997 年 9 月 5 日初稿
1998 年 3 月 4 日修订
1998 年 4 月 17 日修订
1998 年 12 月 10—12 日修订
1999 年 6 月通过

国际建筑师协会职业实践委员会秘书处

美国建筑师学会
联合主任　J.A. 席勒
1735 New York Avenue, NW
Washington, DC 20006
　电　话：202 - 6267315
　传　真：202 - 6267421

中国建筑学会副理事长
联合主任　张钦楠
中国北京西城区百万庄
邮　编：100835
　电　话：8610 - 88082239
　传　真：8610 - 88082222

认同书中关于注册/执照/证书的政策

UIA 促进在世界各国建立建筑师的注册／执照／证书制度。从公众的利益出发，这种制度的条文应当由法律规定。

引　言

注册／执照／证书

注册／执照／证书是官方对某一个人可以作为一名独立的建筑师进行实践的个人资格的法律承认，该法律同时防止不合格的个人履行某种职责。考虑到在一个高质量、可持续发展的建造环境中的公众利益以及与建筑工业相关的危险和后果，建筑服务由适当的、合格的职业人员提供，对于向公众保证足够的保护至关重要。

注册／执照／证书应以有关教育、经验和考试的最低能力标准为基础，以保障公众利益。职业执照是国家为保障公众的健康、安全和福利而行使其被赋予的权力。以下五种准则被普遍接受为适宜于颁发执照的条件：（1）对该职业如不加管理将会对消费者的生命、健康安全或经济福利产生严重风险，其潜在损害是显然和极其可能的；（2）该职业的实践要求有高度的技能、知识和培训；（3）实践者的功能和责任要求独立的判断，而该职业的成员是独立执业的；（4）该职业的实践范围可以和其他有或无执照的职业区别；（5）对该职业实施管理的经济和文化影响对公众来说是合理的。建筑师实践符合以上原则。

实践注册与称号注册

"实践注册"，即对某一职业的实践的管理，意即只有符合专门的立法准则（教育、培训和测试）的个人可以执行该职业的服务。

实践管理或执照——由于其对国家和消费者产生费用、以及它对进入该职业的限制，在传统上国家只限于对那些如不加管理将对公众健康、安全和福利构成严重威胁的职业进行管理。在评价一个职业是否应当实行职业管理时，多数国家应用一组客观的准则，包括：公众是否会由于缺乏管理而受到损害？除国家管理外还有何种选择？公众是否受到现有法律、规范或标准的保护，加强这些法律能否解决问题？职业管理要给国家和公众增加多少负担，公众是否从中受益等。

"称号注册"，意即个人仍需符合特定的资格准则，但只控制称号的使用，没有称号的个人仍可以继续承担服务项目。称号注册只是授予被保护的称号。一项称号法案不影响这组人员的实践范围或允许这些个人去承担法律未允许的工作。

（注：称号注册在多数国家被称为"证书"，而"执照"则多数用于实践注册）。

称号注册能够提供一种手段使公众可以用来区别受过培训的合格实践人员或服务提供人与未受培训或不合格的个人，对国家和消费者发生的费用较低。在称号注册制度下，那些不符合注册要求的个人也没有被取消服务资格，他们可以继续提供服务，但不能使用受保护的称号。

称号注册使公众和服务的消费者以最少的国家和消费者花费，就能区别受过训练的合格人员和没有受过训练的人员。采用称号注册制并不剥夺那些未符合注册要求人员的生计，这些人员能继续提供服务，他们仅限于做那些使用无需受保护称号的工作。

提议的立法导则

UIA 推荐管理建筑师职业的立法或条例应以实践注册为基础。以下导则反映了此项推荐，它提出的条款所涉及的问题可能超出了国界范围的含义。为简要起见，以下均以"注册"二字代表"注册/执照/证书"。应注意在任何国家或国际机构之间的相互承认协议中，UIA 的立场是：只承认注册建筑师（不论是实践注册或称号注册）。由于 UIA 成员国组织所代表的国家(地区)在语言、组织和条例上反映了各自的政治和文化特征,因而本导则只提供导则性的文字而不用法律语言。在国际基础上使用严格的法律语言会导致许多争论和混乱。

1. 定义

1.1 建筑师实践:在注册立法中,建筑师实践的定义应采用《UIA关于建筑师实践中职业主义的推荐国际标准认同书》中的提法。

建筑师实践包括提供有关城镇规划以及一栋或一群建筑的设计、建造、扩建、保护、重建或改建设计的职业服务。这些服务包括但不限于规划、土地使用战略及规划、城市设计、提供前期研究、设计、模型、图纸、说明书及技术文件,对其他专业(咨询工程师、城市规划师、景观建筑师和其他专业咨询师等)编制的技术文件作适当的协调以及提供建筑经济、合同管理、施工监督与项目管理等服务。

建筑师实践的定义覆盖了广阔的服务范围,建筑师一般均在这些方面受到专门训练,并在实践中要求其显示自己的职业能力。在某些地区,教育和培训的范围窄于以上者,对UIA认同书中的定义可相应调整。

除已注册或按照注册条例被允许实践者,任何人不得从事建筑师实践。不是注册建筑师的任何人(他/她)不得使用"建筑师"称号或向公众以及其他方式表示他或她是一位建筑师。

在某些地方,法规可能没有将其他设计专业的范围规定清楚,使其他设计专业也可附带性地从事建筑学的实践。重要的是,要慎重地将其他合法的设计活动和建筑学的实践区分开。

在许多地方,工程注册法允许工程师设计结构物和许多其他项目,但建筑师职业通常由法律规定只设计房屋建筑和"为人居环境"的附属设施。UIA主张对建筑师职业的法律不应有不必要地限制实践范围,并应认识建筑师通过其实践,在自己的建筑设计中表达了对社会的文化的根和美学价值。

2. 注册人的行为管理

2.1 授权:显然,应当由法律授权建筑师注册机构制定管理建筑师行为的规定。制定规定的权力以及由于不当行为而取消或暂停注册的权力就自然要求在制定法规时要进一步描述不当行为的构成。

2.2 行为法则：法规应当授权建筑师注册机构在其管理功能中颁布注册建筑师实践行为条例。法规应当包括这些条例的范围和内容方面的标准。法规还应当确定违反由建筑师注册机构所制定的行为法则将导致注册的取消或暂停，或民事罚款。

3. 注册资格

资格标准应当客观和透明。在法律条款中，应注意能反映《UIA关于建筑师实践中职业主义的推荐国际标准认同书》中对一名建筑师的基本要求、教育、评估、实践经验和考试等的政策和导则。在法律中，不应当提出进入建筑师职业要有公民或居住资格等条款。

3.1 学位：注册申请人应持有经过评估合格的建筑学学位，以及由注册机构根据其规定认为是合格的实践培训。UIA推荐以UIA/UNESCO《建筑教育宪章》作为建筑教育的最低标准。

3.2 实践训练：UIA推荐，注册申请人具有"UIA认同书"中所规定的实践训练。

3.3 考试：注册申请人要以注册机构的规定为基础，通过有关科目的考试及评分。

3.4 个别面试：注册机构可要求对注册申请人进行一次面试。

3.5 道德标准：为慎重起见，注册机构可以根据不具备"良好道德品质"为由拒绝一名申请人，法律应当具体规定有关的调查内容，例如：

• 有犯罪行为而被判处者；
• 在申请中有虚报假报者；
• 违反由法律或规程所规定的行为法则者；
• 违反本辖区注册法，未经注册就从事建筑师实践者。

如果申请人的背景包括以上内容，注册机构可根据申请人改过的情况，允予注册。

4. 对等程序

除了在第三节（注册资格）以及其他与对等有关的条款外，法律还应当对非居民申请注册作出规定。

4.1 非居民申请注册：每个寻求在某辖区实践的非居民均应注册，其条件是申请人：

• 持有由该辖区通过相互承认协议所承认的注册机构所颁发的现行有效的注册；

• 按照本辖区规定的报表格式提出申请，包括有关该辖区认为需要的能令该辖区满意的信息。

4.2 非居民寻求委托：一位非居民在他／她未注册的辖区内寻求建筑设计委托者，如符合下列条件，可以允许在尚未注册的情况下提供建筑服务：

• 持有由该辖区通过相互承认协议所承认的注册机构所颁发的现行有效的注册；

• 用书面方式向辖区的委员会通报：(a)他／她持有由该辖区通过相互承认协议所承认的注册机构所颁发的现行有效的注册，但尚未在本辖区注册，将要争取在本辖区提供建筑服务；(b)他／她将向自己要争取服务的业主提交(a)中的通报副本；(c)他／她保证一旦被选中为该辖区的建筑师，就立即向辖区委员会申请注册。

4.3 设计竞赛：一位通过在他／她没有注册的辖区参与建筑设计竞赛、寻求委托者，应当被允许参与竞赛，其条件是：

• 持有由该辖区通过相互承认协议所承认的注册机构所颁发的现行有效的注册；

• 用书面方式通报该辖区他／她参与设计竞赛、并持有由该辖区通过相互承认协议所承认的注册机构所颁发的现行有效的注册；

• 一旦被选为该项目的建筑师，就立即向辖区申请注册。

5. 实践形式

当建筑服务由公司实体提供时，要求公司实体要在建筑师的有效控制之下，并与个体建筑师一样符合有关服务、工作和行为方面的职业标准。

多数回答 UIA 职业实践委员会调查表的成员组织指出其所在

国允许以合伙制或常规公司制方式进行建筑师实践。对公司制实践及新的有限责任制企业的限制往往偏于苛刻。这种多样的限制说明，需要有一个导则可以就公司制实践寻求一种合理的国际性的规定，同时又能保证公众取得建筑师服务的完整性。

5.1 实践结构：UIA 导则推荐在一辖区的法律下应允许以合伙制（包括注册的有限责任合伙制）、有限责任公司或公司的形式进行建筑师实践，只要满足以下要求：

• 在合伙制中有三分之二的合伙人；在有限责任公司或大公司中，有三分之二的董事已按照国家或州的法律注册并进行建筑实践；

• 在合伙制中负责建筑师实践的人为主要（general 也可译成"总"）合伙人；在有限责任公司或大型公司中，为一董事。以上所列人员均应按照国家或州的法律在本辖区内注册。法规应授权注册机构可以要求所有合伙人，董事，股东及其经营管理层的有关信息。（本段的意思是有的公司并不是单做建筑设计，如也做结构、设备设计，或者是房地产，或者是施工，或者是像日本的"三菱"等什么都做，但只要管建筑设计的是各注册建筑师即可。不过对这类公司要提供公司其他领导层的信息。——译者注）

5.2 企业名称：一个在其他方面均符合在某个州或国家实践的单位，应允许采用一个并不包括所有已在本州或本国注册的公司董事、有限公司董事或合伙人名（在合伙制）的名称，只要该企业已按注册机构的合理要求申报其董事或合伙人的有关信息。（这一条是指一个公司在其他地方设点时的情况。——译者注）

6. 建筑施工期间建筑师的雇佣

施工期间建筑师的施工管理服务包括定期现场访问、加工图审核、违反规范或与合同文件的重大偏离的报告等，建筑师的重要责任是保障公众的健康、安全和福利。下列导则旨在保证设计建筑师最低程度的施工服务。

6.1 一位正在建造一项主要是供人们使用或居住项目的业主，如果没有雇佣一名建筑师来完成最低程度的施工管理服务，包括定期现场访问、加工图审核以及向业主及建筑官员报告任何违反规范或与合同文件有重大偏离的情况等，应当视为自己在从事建筑实践。

6.2 如果项目设计建筑师没有被雇佣于上述第一段中所述的施工管理服务时，他／她有责任向注册机构以及建筑官员报告。

6.3 如果注册机构确认即使没有如上述第一段中所述的由建筑师完成的服务而公众仍能得到足够的保护，就可以对某一特定项目或某类项目免除此类要求。

7．对未注册人员进行建筑实践的规定

未注册的建筑实践会危害公众的健康、安全和福利。以下导则提供了强制实施法律的基础和途径。

7.1 未注册人员违反建筑注册法规应属犯罪行为，注册机构应被授权，在一次听证后，可以给予一定限额的民事罚款，以及发布命令制止未注册人员及协助或纵容未注册人员的行为。注册机构以及政府的监察官或其他地方执法单位，应被授权可以禁令未注册人员及协助或纵容非注册人员者的行为，注册机构并享有给予民事罚款的权力。

7.2 所有由地方建筑管理部门规定要归档的建筑师实践（在导则一中描述者）中提供的图纸、说明书和其他技术文件，均应有一名建筑师的印章。假如法律在一般要求之外允许有例外时，提交该技术文件者应具体指出该法律条款。任何不符合以上的文件均属无效。

国际建筑师协会
建筑实践中职业主义的推荐国际标准

关于道德和行为标准的政策推荐导则

1997 年 10 月 31 日初稿
1998 年 4 月 17 日修订
1998 年 12 月 10—12 日修订
1999 年 6 月通过

～～～～～～～～～～～～～～～～～～～～～～～～～～～～

国际建筑师协会职业实践委员会秘书处

美国建筑师学会
联合主任 J.A.席勒
1735 New York Avenue, NW
Washington, DC 20006
电　话：202-6267315
传　真：202-6267421

中国建筑学会副理事长
联合主任　张钦楠
中国北京西城区百万庄
邮　编：100835
电　话：8610-88082239
传　真：8610-88082222

认同书中关于道德和行为的政策

现有的 UIA 关于咨询服务的国际道德规范继续有效。鼓励 UIA 成员组织在自己的道德和行为规范中引入和实施一项其成员应按遵守在其提供职业服务的国家和辖区现行的道德和行为规范的要求，只要该规范的规定不受国际法或建筑师本国法律的禁止。

引　言

在 1998 年 12 月于华盛顿召开的委员会会议上，有广泛的一致意见赞同把产生于巴塞罗那会议的本规范提交北京的代表大会，采纳作为认同书关于道德和行为的导则，随后由各成员组织采纳于其本身的规范中。起草小组根据认同书中阐明的原则与政策，以及世界各国成员组织的道德与行为规范，向理事会和代表处推荐如下。

序　言

建筑师职业的成员应当恪守职业精神、品质和能力的标准，向社会贡献自己为改善建筑环境以及社会福利与文化所不可缺少的专门和独特的知识和技能。以下原则是建筑师在完成其咨询服务任务时的行为原则。它们适用于所有的职业活动，不论在何处出现。它们涉及对本职业所服务和造福的公众的责任；对业主、建筑用户和帮助形成建造环境的建筑业的责任；对建筑艺术和科学的责任，这种知识和创造的延续体已成为本职业的遗产。

原则 1
总的义务
建筑师通过教育、培训和经验取得系统的知识、才能和理论。

建筑教育、培训和考试的过程向公众保证了当一名建筑师被（聘用于）指定完成职业服务时，该建筑师已符合为适当完成该项服务的合格标准。建筑师有总的义务，要具有并提高其建筑艺术和科学的知识，尊重建筑学的集体成就，并在对建筑艺术和科学的追求中把以学术为基础和不妥协的职业判断置于其他各种动机之前。

1.1 道德标准：建筑师要努力提高其职业知识和技能，并在涉及其实践的领域内维持其职业能力。

1.2 道德标准：建筑师要不断寻求提高美学、建筑教育、研究、培训和实践的标准。

1.3 道德标准：建筑师要尽可能地推进相关行业，并为建筑业的知识和能力作出贡献。

1.4 道德标准：建筑师要保证其实践有恰当和有效的内部程序，包括控制和审查程序，并有足够的、合格的、处于监督之下的工作班子，使他们能有效地进行运作。

1.5 道德标准：当某项工作由一名雇员或任何其他人代表建筑师在其直接控制下完成时，建筑师有责任保证此人有能力完成该任务，并在需要时处于充分的监督之下。

原则 2
对公众的义务

建筑师有责任从精神到条文遵守管辖其职业事务的法律，并周到地考虑到其职业活动所产生的社会和环境影响。

2.1 道德标准：建筑师要尊重和保护他/她们所执行任务的社区的自然和文化遗产，同时又要努力改善其环境和生活质量，注意到其工作对将要使用或享受其产品的所有人的物质和文化权益。

2.2 道德标准：建筑师在其职业服务中不宜以虚假、误导或欺骗的方式进行沟通或自我推销。

2.3 道德标准：建筑师实践的营业风格不宜产生误导，例如造成其他实践或服务的混乱。

2.4 道德标准：建筑师在其职业活动中要遵守有关的法律。

2.5 道德标准：建筑师要遵守他/她所提供职业服务的国家及

辖区现行的道德与行为规范，只要这些规范没有受到国际条约、协定和法律以及建筑师本国法律的禁止。

2.6 道德标准：建筑师要作为公民和职业人士适当地参与公共活动，并促进公众对建筑问题的了解。

原则 3
对业主的义务

建筑师应忠诚地、自觉地对业主承担义务，以职业方式执行其职业工作。在所有的职业服务中，合理地考虑到有关的技术和职业标准，作出无成见和偏见的判断。在建筑学的艺术和科学追求中，学术性和职业性的判断应当比其他任何动机处于优先地位。

3.1 道德标准：建筑师只有在确定可有足够的财务和技术资源提供的条件下使其能全面完成自己对业主所承担的义务时才可承接职业工作。

3.2 道德标准：建筑师要以应有的技能、关注和勤劳来完成其职业任务。

3.3 道德标准：建筑师要不拖延地，并在其力所能及的限度内在合理的时间内完成其任务。

3.4 道德标准：建筑师要及时告知业主有关他/她代表业主所承担的工作的进展，以及影响质量和成本的各项问题。

3.5 道德标准：建筑师要对自己向业主作出的独立的意见承担责任。建筑师和他们所聘用的咨询师只承接他们经过教育、培训和经验，在专门领域中合格的职业服务工作。

3.6 道德标准：建筑师在业主方以书面方式清楚地写明委托条件之前不宜承担职业工作，特别是：

- 工作范围
- 责任分工
- 责任限制
- 收费数量及方式
- 终止条件

3.7 标准：建筑师的酬劳、设计费用，应在聘用合同中详细写明。

3.8 标准:建筑师不应该有任何获取好处的动机。

3.9 道德标准:建筑师应注意其业主事务的保密性,在没有取得业主或其他法律权威(例如,根据法庭的命令的透露)的事先同意下,不宜透露保密的信息。

3.10 道德标准:建筑师要向业主、物主或承包商告示任何他/她所了解的可以被解释为产生利益冲突的重要情况,并保证这种冲突不会损害这些方面的合法利益或构成对建筑师公正判断其他方面执行合同情况的责任的干预。

原则 4
对职业的义务

建筑师有义务维护本职业的品质与尊严,在所有情况下都以尊重他人的合法权利和利益的方式行动。

4.1 道德标准:建筑师要诚实和公正地从事其职业活动。

4.2 道德标准:建筑师不宜将一位不适宜的人吸收为合伙人或经理,例如,已从注册名单中除名者(自己要求退出者除外)或已被公认的建筑师组织排除者。

4.3 道德标准:建筑师要努力通过其行动促进其职业的尊严和品质,并保证其代表和雇员都按能符合此标准,以免由其行动或行为使其所服务者或共同工作者的信任受到颠覆,或使与建筑师有关的公众不受曲解,欺骗和伪造的损害。

4.4 道德标准:建筑师应该尽力地、努力地为建筑学知识、文化和教育作贡献。

原则 5
对同行的义务

建筑师要尊重其同行的权利,并承认其同行的职业期望、贡献和工作成果。

5.1 道德标准:建筑师不宜有种族、宗教、健康、婚姻和性别上的歧视。

5.2 道德标准：建筑师在取得原来建筑师的明确授权之前，不宜采用其设计概念。

5.3 道德标准：建筑师在作为独立咨询师提供服务时，或未被要求报价之前，不宜提出报价。必须要准备有足够的有关工程范围的自然环境信息才能报价，其中要明确指出费用涵盖的服务内容。

5.4 道德标准：建筑师在作为独立咨询师提供服务时，不宜因考虑其他建筑师对同一服务的报价而修改自己的报价。

5.5 道德标准：建筑师不宜以不公正的手段试图挖取另一建筑师已取得的委托。

5.6 道德标准：建筑师不宜参加 UIA 或其成员组织宣布为不能接受的建筑竞赛。

5.7 道德标准：建筑师不宜担任竞赛评定人，而又以另一身份承接该工作。

5.8 道德标准：建筑师不宜恶意地或不公正地批评或试图毁谤另一建筑师的工作。

5.9 道德标准：建筑师在被接触承担一个项目或其他职业任务，他/她已知道或经过合理询问后可确定另一建筑师已由同一业主处取得同一项目或职业工作的委托时，应当把事实告知该另一建筑师。

5.10 道德标准：建筑师被指定要对另一建筑师的工作发表意见时，应将此事告知该建筑师，除非有证据这样做会对将来可能发生或已经发生的诉讼产生偏见的情况，否则这样做会认为是种偏见，会引发将来或现在的诉讼。

5.11 道德标准：建筑师要给其助手和雇员提供适宜的工作环境，给予公正的报酬，并便利其职业发展。

5.12 道德标准：建筑师要保证其个人和职业的财务得到慎重的管理。

5.13 道德标准：建筑师要把自己的职业荣誉放在其自身服务和成就的基础上，对他人完成的职业工作的成就要给予承认。

国际建筑师协会
建筑实践中职业主义的推荐国际标准

关于继续职业发展的政策推荐导则

1997 年 10 月 31 日初稿
1998 年 3 月 11 日修订
1998 年 4 月 17 日修订
1998 年 12 月 10—12 日修订
1999 年 7 月通过

~~~~~~~~~~~~~~~~~~~~~~~~~~~~~~~~

国际建筑师协会职业实践委员会秘书处

美国建筑师学会
联合主任　J.A. 席勒
1735 New York Avenue, NW
Washington, DC 20006
电　话：202 - 6267315
传　真：202 - 6267421

中国建筑学会副理事长
联合主任　张钦楠
中国北京西城区百万庄
邮　编：100835
电　话：8610 - 88082239
传　真：8610 - 88082222

认同书中关于继续职业发展的政策

UIA 及其成员组织从公众利益出发,鼓励继续职业发展,这是 UIA 成员的责任。建筑师应保证他们能够提供合格的服务,行为规范要求建筑师应保持"基本要求"及其未来变化中所列的各种范围内所要求的标准。同时,UIA 将监督以继续职业发展作为更新注册条件以及这一做法的发展。推荐有关导则以促进国家间的对等承认,继续完善此项政策。

导 则

继续职业发展不同于为取得更高学位的教育,而是指整个一生的学习过程,以维持、改进、或增添建筑师的知识和技能,以确保其知识和能力能适应社会的需要。

UIA 的政策是鼓励其成员组织把继续职业发展视为每一位建筑师的责任。继续职业发展也对公众有利。

本导则旨在向 UIA 成员提供一套标准,可以对其现行继续职业发展政策进行判断。它将保证政策的适宜性并为将来的对等承认和职业发展的学时的转移创造条件。

本导则的最初目标是为在 UIA 成员组织之间相互承认继续职业发展建立框架。

UIA 成员组织的继续职业发展的关键要素为:

• 推荐关于确认、审查、评估继续职业发展的服务和课程的程序;

• 推荐关于自学课程要求和继续职业发展机构所提供课程的要求;

• 推荐将研究和设计需求相结合的继续教育授课准则;

• 推荐确保将学习的重点放在学习者和知识的获得上的过程,包括鼓励在继续教育提供者和接受者之间有互动的课程,使有互动课程的学分高于同一时间内无互动课程的学分；
- 保证教育的质量，而不是只看时间；
- 建立记录制度，使得继续教育提供者能从参与者那儿取得反馈；该记录本可以在世界范围内实现学时转换之用，以及供那些要求有继续职业发展作为注册或保持会员资格的条件时所用；
- 提出每年要求的学时；
- 提出对保护公众健康、安全和福利内容的最低学时要求。

UIA成员组织的继续职业发展系统应确立高水平的教育标准，有大量肯传授其知识、技能和研究的在册专家（提供者），努力取得成功。

国际建筑师协会
建筑实践中职业主义的推荐国际标准

关于在东道主国家建筑实践的政策推荐导则

2002年2月26日—28日西班牙巴塞罗那节95届理事会通过
2002年7月27日—29日德国柏林第22届代表大会通过

~~~~~~

国际建筑师协会职业实践纲要联合秘书处

美国建筑师学会	中国建筑学会副理事长
联合主任　J·席勒·FAIA	联合主任　许安之
1735 New York Avenue, NW Engineering	中国北京百万庄
Washington, DC20006	邮　编：100835
电　话：202-6267315	电　话：8610-88082239
传　真：202-6267421	传　真：8610-88082222

认同书中关于在东道主国家实践的政策

建筑师在他们没有注册的国家提供一个项目的建筑服务时，应当与当地的建筑师合作以保证恰当和有效地理解当地法律、环境、社会、文化和传统方面的因素。合作的条件应由各方依据国际建筑师协会的道德标准和当地的法律法规确定。

关于在东道主国家实践的政策推荐导则

序　言

国际建筑师协会（UIA）鼓励在《关贸总协定》(即 GATT）和《服务贸易总协定》(即 GATS）及世界贸易组织的框架内达成双边或多边协议。在建筑师职业互认或自由贸易谈判过程中的经验建议，双方要明确外方的职业资质标准与当地的职业资质标准间的差别，双方应谈判如何通过设立等效的资质标准来沟通这种差别。在互认谈判过程中，必须认可各管辖区有制定它们自己职业标准的主权，但所有职业标准都应反映出建筑师对环境、社会、文化、公共健康，公民的安全和福利的责任。

认同书认为 UIA 各成员组织的标准，其实践和状况存在着差别，反映了各国文化的多样化。认同书是国际建筑师的代表们对于这些标准和实践达成共识努力的第一步。UIA 认识到双边和多边的互认和/或自由贸易协定可能需要时间进行谈判和付之实施，因此需要提供专项的导则和草案给尚未互认和/或自由贸易协定的地区和国家。

认同书中关于在东道主国家实践的政策认为，各相关国家建筑师之间的地位是平等的，本导则意图是为现在各国互认和/或自由贸易区协定已经普遍存在的今天起到一个桥梁作用。下面的导则建议了一份推荐条款供 UIA 成员组织采纳，而这些成员组织也在探讨合适的机制以认可外国建筑师于东道主国的实践。

引 言

在大多数有法律管辖权的地区和国家，建筑师从事建筑实践必须注册，有执照，或有证书。本导则适用于建筑师个人或建筑师的公司在他们未有注册、执照和证书的国家受委托做设计。

UIA 认识到，存在着对建筑师可流动及在外方管辖区提供服务的需求，这也是 UIA 的一个目标，使任何一位 UIA 成员组织中被有关当局认可的建筑师，也被所有 UIA 成员组织所在的国家和地区通过双边或多边协定被有关当局认可。

UIA 还认识到，需要增强对当地地方环境、社会、文化因素和道德及法律标准的意识。到目前为止，UIA 代表大会已经通过了"UIA 建筑实践中职业主义推荐国际标准认同书"（第二版），认同书及其相关的政策导则是想为建筑师职业追求的标准达到最佳实践，并想使有关的各方在双边互认及自由贸易协定谈判时和建筑师的资质和服务谈判时比较容易。

早已制定的 UIA 国际咨询服务道德规范要求："每一个从外国来的顾问（建筑师）应与项目当地的顾问或专业人士协作一道和谐地工作。"

在东道主国家实践的导则

UIA 推荐采纳本导则的各成员组织（国家和地区）同意以"UIA 认同书"中的政策框架来谈判当地建筑师和外国建筑师的合作协定。

UIA 认同书及其相关的导则意在建立国际建筑实践标准，同时也要认识到，UIA 各成员国的传统和实践存在着差别。

建筑师加入"东道主国家实践"协定应同意：1)与职业责任、保险、法律的辖区及诸如此类的相关问题应由当地法规作出规定或由有关的商务商定。最好是由当地建筑师、外国建筑师和业主一起谈判并正式写进双方或多方的协议中；2)影响建筑师建筑实践和行为的公共法定责任、法规，法律应由双方评估鉴定，它是双方建筑师共同的责任；3)下列状况适用于外国建筑师在当地管辖区的建筑实践：

1．本导则中所述的建筑师是一位具有职业资格并在相关地区和国家当局注册/有执照/有证书的建筑师。当地建筑师是已注册/有执照/有证书的实体，并在项目所在国家从事建筑实践。外国建筑师是已注册/有执照/有证书的实体，并正在一管辖区和国家实践，但没有在项目所在的辖区注册/执照/证书。

2．在东道主国家有关当局和外国建筑师所在国双方没有互认或自由贸易协定的地区和国家。

• 在他们本国有关当局注册/有执照/有证书的外国建筑师，但未在东道主国家注册，应当个别地予以认可，并允许其与当地注册/有执照/有证书的建筑师按照当地的法律和实践进行合作。

• 来自没有相关管理注册/执照/证书的国家和地区的外国建筑师，应当按项目所在国家和地区所规定的有效标准，取得其注册/执照/证书资格。

• 外国建筑师如没有当地建筑师的实质性参与就不应让他在当地提供服务。本条款不适用于国际设计竞赛中符合资格的建筑师的方案提交。当外国建筑师被选中后，应与当地建筑师合作。

2.1 外国建筑师应当：

a．准备证明材料给有关国家或国际当局，证明其持有一个管辖区有关当局发的有效的注册证/执照/证书，这些注册证/执照/证书允许他们使用"建筑师"头衔并可在其所在的辖区从事无限制的建筑实践。

b．提供他们的资格证明。

c．提供他们无任何刑事犯罪和道德犯罪的证明。

2.2 外国建筑师被选定为项目的设计者之后，应立即要求当地建筑师向有关当局提供一份文件，文件中规定当地建筑师和外国建筑师的关系，包括他们的资格证明，以及一份对于该合作设计项目的说明。

2.3 外国建筑师和当地建筑师都应确信他们在合作中的双方

都具有该项目所需要的技能和经验。

2.4 项目合作双方建筑师提供的职业服务应当有联合服务和各自地进行的服务。

2.5 在项目任何设计文件中,当地建筑师和外国建筑师都应准确地显示他们对项目各自的责任。

2.6 根据地方有关当局要求,UIA成员组织所在国的有关当局将同意回复确认该外国建筑师的身份资格,如2.1条款所建议的那样。

2.7 应要求外国建筑师同意遵守所在管辖区的法律、道德和行为法规、建筑法规等。

2.8 外国建筑师和当地建筑师形成合作关系时应要求他们签订正式的、公平的、对等的协议,协议要体现UIA的道德标准。关于合作建筑师之间签订的协议,已经有许多文件和书提供了范本,涵盖和考虑了上述原则和问题。

3. 本导则不适用于两个相关国家和地区已有互认协议的情况下。

《在东道主国家实践》起草小组

J·席勒(James A.Schecler,组长)

L·罗西(Luis M. Rossi)

C·费耶特(Carlos Maximiliano B.Fayet)

A·汉姆普尔(Andreas Gottlied Hempel)

T·普林(Tillman Prinz)

E·波汉默德(Dat'OHahi Esa Bin Bohamed)

A·法辛斯基(Artur Fasinski)

E·西尔伐(Edward D'Silva)

S·爱伦(Susan M.Allen)

国际建筑师协会
建筑实践中职业主义的推荐国际标准

关于知识产权和版权的政策推荐导则

2002 年 2 月 26—28 日　西班牙巴塞罗那第 95 届理事会通过

UIA 职业实践委员会联合书记处

美国建筑师学会
联合主任　詹姆士·A·席勒 FAIA
1735 New York Avenue, NW
Washington, DC2006
电　话：202-6267315
传　真：202-6267421

中国建筑学会副理事长
联合主任　许安之
中国北京　西城区百万庄
邮　编：100835
电　话：8610-88082239
传　真：8610-88082222

认同书中关于知识产权和版权的政策

UIA 成员组织的国家法律应当在无损于建筑师权责的条件下给从事职业实践的建筑师有权持有他（她）的作品的知识产权和版权。

UIA 希望尽可能有效和始终如一地发展和保持对建筑师作品的知识产权及版权的保护，UIA 认识到信息量的集聚和信息技术的发展对艺术作品的使用产生巨大冲击，认识到需要平衡作者权利和公众利益的同时，还要强调版权保护的重大意义以激励艺术创作。

本导则所提到的作品应在全体 UIA 成员组织所在的国家受到保护。这种保护将有利于作者本人以及他（她）的合法继承人。

序　言

建筑师服务的特点是表现了建筑师的智慧和能力。合格的建筑师在考虑业主要求时能运用知识和技能做出具有创造性的建筑，建筑师智力活动所创造的想法和概念是一种产品，使建筑师能去实施建造。这种创造性的智力努力要求有很好的保护。知识产权保护保证其他人不能使用这种建筑师们的智力努力，也不能使用其他原创性的作品。这类保护对于建筑的进一步创新发展是一种鼓励。它将对业主和公众有利。它对于建筑中文化的提升很重要，为了能使人们识别在建造环境中的自身，建筑文化在全球化的世界中已变得越来越重要。

知识产权是指工业、科学、文学及艺术领域中的智力活动所导致的权利，如版权、专利权等等。相对物品的产权而言，知识产权指的是智力的努力。知识产权是一个总称，具体一点就有专利法、版权法和贸易商标法。版权特别包含了与艺术创作有关的智力努力，而与纯技术过程和目标有关的智力努力在专利法中受到保护。

版权保护了与作者创作有关的权益,并授予他们有使用他们创作的专用权。

为了使建筑师在外国也能成功地提供服务,必须保证其创作作品的知识产权受到保护。因此,知识产权在所有国家受到保护很重要,这样才能使建筑师有信心对他们的业主提供最好的服务,在这方面,UIA 关于东道国实践的政策推荐导则对于跨境服务时建筑师知识产权的保护起到了重要作用。

导 则

下列导则的目的是确定主要概念和问题及通常可能发生的建筑领域内的知识产权,包括与版权有关的"作者"和"作品"的定义以及诸如道德标准权、保护、所有权及实施等问题。

1. 作者

1.1 初始拥有权

一件作品的作者是创造该作品的个人,因此作者推定为该作品版权的所有者。作者永远是一位自然人。如作品产生在顾问合同条件下(见 1.2)或者作者同意转让他们的版权,只要相关的国家版权法中有此类规定,则公司、商业机构和公共事业的实体也可以拥有版权。

1.2 在雇用合同和顾问合同条件下的作品

一位建筑师在雇用合同下所作的作品,可推定雇主是版权的所有者,但如果在雇用合同中有明确规定的除外。如果一位建筑师作为顾问所做的作品,该建筑师可推定为是版权的所有者。但如法律允许,版权可以根据作者同意的协议转让,由于版权是一种商业财产,因此在转让时除了建筑服务付费以外还要另付转让费给同意转让作品版权的建筑师。

1.3 集体作品

一项集体作品如由各独立的人完成,在最初阶段就应明确职责分工,这种情况下其版权与一项整体的集体作品是很不同的。如果没有明确的版权和任何权益转让，这类集体作品版权所有者可以推定具有复制和分发自己所完成的那部分权利，任何集体作品的修改版，以及后来的集体作品也以此类推。

一项由集体完成创作作品的版权由该项作品的全体作者所有。所有作者对版权具有同等权利，就作品整体而言，只能集体使用这种权利。

2. 受保护的作品

版权保护"作者的原作"，原作有固定的有形的表达形式。底片等不能直接感觉得到，但只要借助于一台机器或装置就能交流。作品一旦被创造出来其版权的保护就自动生效，无需出版、注册或其他行动。

2.1 建筑作品

本导则所保护的作品是作者的原创建筑作品，固定在任何有形的表达方式上，它代表了个人原创性的智力创作。新颖、独特和美观的水平不是被保护作品的评判标准，但原创性是必要标准，要求作品是作者努力的结果而不是一个现有作品的抄袭。

版权保护范围延伸至文件形式或建成的作品，但不包括构思、程序、实施的方法及数学概念，因为这些工作可以在相关的技术保护如专利权中得到保护。版权保护可以覆盖任何类型的建筑作品。

2.2 特殊建筑作品的保护

2.2.1 建筑设计文件

建筑设计和建造文件无论是电子文件还是图纸形式都受版权保护。除了项目的图纸、草图等受版权保护外，如项目已实施为三维，项目本身也能受到保护。这也包括了城市规划和城市设计。

2.2.2 专家意见、说明书及其他文件

已成文的专家意见、说明书及其他文件也能受到版权保护，只要它们代表了个人的创造。这种保护不包括文件的内容而仅限于提出的形式，目的是为了将版权与技术保护，如专利权加以区分。

2.2.3 建筑物

一座建筑能得到版权保护，只要设计体现了个性化原创性的要求（见 2.1）。同样，建筑的局部或多个建筑结合的建筑群及原建筑群中与已有元素结合的新建筑创作都可受到版权保护。风格、品位、美学价值及时尚等因素并不重要，只要作品具有创造特色，任何建筑物和建筑作品都可受版权保护。

3. 作者利益的保护

建筑师作为版权的拥有者有对他作品的版本进行复制的专有权，只要这种复制在版权法的保护之下。作者有权对他人的未经授权的复制采取法律行动。但在建筑领域，必须注意到许多建筑元素已广为人知，如门、窗、屋顶、墙，并因此限制了建筑创作。这就是为什么如果一座建筑物的独特概念、特别的技术细部及一个建筑的独特的外形被模仿时，建筑版权受侵犯就增加。现有建筑物对新设计的建筑作品仅有影响并不构成对版权的侵犯。

享有权利的道德标准／有道德标准的权利

被称之为道德权利包括了受保护作品的归属权和整体权。这些权利提供了在受侵犯时，对作者的确认及对作品的保护。

3.1 建筑作品的出版权

与建筑作品作者相关的法律应给予作者有出版他们作品的专有权。这种专有权是有限制的，因为通常建筑师不好对业主的建筑出版发表意见。然而，任何时候建筑师都应当有权决定出版其平面和建筑照片。另外，建筑设计竞赛的方案只能按竞赛规定条件或作者的同意才可以出版或展览。

学生在设计课程中的作业仅在学科评估和评论需要时才能出版或展览。别的情况下没有学生的同意不得出版学生作业。

作为评论、新闻、报道、教学、奖学金或研究等目的使用有版权的作品时不应侵犯版权。这些用途的使用有时称为"公平使用"，只要不侵犯版权，这类公平使用有一定限度，因为它会减少作者的版权作品的市场。

3.2 著作权的确认

作者应当有权在其作品上落下他们的名字，有权在作品出版时署名。这种确认对于草图、平面和其他文件尤其重要，它也适用于建筑本身。即使一个版权的注识标记并不是版权保护的必要条件，但版权标记应当表示出来，特别是在图纸和其他建筑文件上应清楚标志。这样做，作者可防止不经意的侵犯，减轻实质上或法令上的损害。版权注识可以是："版权所有$^{(C)}$某某建筑师事务所，1999"。

3.3 建筑作品的侵犯

除了作者的经济权外，甚至即使经济权已转给别人，作者仍然有权拥有作品的署名权，有权反对抄袭、损坏、复制及修改等其他有损于作者荣誉和声誉的行为。这种权利有时被总称为创作中的道德权（名誉权），它应当被维护，甚至在作者去世之后也应维护。在经济权已经失效之后，应当由该国被授权的个人或机构来行使这种道德权。

3.4 改建——当建筑要改建时应平衡建筑拥有者和建筑师的利益

在长期使用中建筑可能进行必要的改建、扩建及其他改变。业主一旦投资于该建筑就必然有可能按经济需求进行改建，业主或建筑使用者有权改变用途，因此，常常包括了建筑构思的变化，有时公共建筑规范的改变也可能要求建筑改动。

在这同时，建筑师的声誉多半已经与该建筑有关。因此，建筑的改动有可能损害已广为人知的该建筑建筑师声誉，改建须保证该建筑师的利益，使建筑具有连贯性，对公众来说建筑不应贬值，水平不应降低。已公开出版的建筑物在未授权的改建后，原建筑师仍然会被认为是该建筑的建筑师，在公众的眼里会认为是原作者设计了改动的部分，因此该建筑师名誉可能受损。

由上所知，有必要平衡改建的业主利益和原设计人的利益。为寻求平衡，必须考虑到将建筑改变用途时该建筑的原建筑师具有更好的见解，因为他对他的创作更了解，所以由他本人来进行改进是最合适的。因此，推荐原建筑师来改进应具有法律上的优先权。这种权利并不禁止业主对改建的要求，但这样的结果使得对业主保持原设计的一致性和对建筑师与改进后的建筑有联系提供了一个可能。

3.5 拆除

有权反对作品的篡改还应包括有权反对作品拆毁。与改建不一样的是，拆除并不像改建那样会构成对原建筑师类似的损害。但拆除仍会对该作品的整体名誉权产生损害，作者的职业声誉与此作品有关。因此，拆除该建筑与该建筑的名誉权冲突。所以建筑物的拆除要考虑建筑师的这一权益。

4. 保护年限

版权保护年限应延伸至作者去世后 50 年。

5. 版权的实施

UIA 建议，应当提供法律和实施程序才能对侵害本导则所包括的知识产权采取有效的行动，这种程序应当避免对合法贸易造成壁垒，并防止滥用。对于知识产权实施的程序应当公平和平等，它们不应当被复杂化或费时、费钱或拖延。

6. 建筑平面的所有权

对建筑平面所有权拥有者的法律规定，在采用通用法的国家和采用拿破仑法典的国家是不同的，在采用通用法的国家，建筑师的设计文件通常被认为是服务的工具，建筑师保留所有权，而业主根据合同可使用执照（资质）和设计文件来建造项目。在采用拿破仑法典的国家，建筑师的文件在合同期满以后成了业主的财产，根据合同，建筑师有义务向业主交出设计文件，这种情况会影响知识产权。在实行通用法的国家，建筑师既是设计文件的拥有者也是该文件知识产权的拥有者。在采用拿破仑法典的国家，建筑师只是文件知识产权的所有者，但具体的文件所有者是业主。

7. UIA 成员组织之间的合作

一个 UIA 成员组织进入另一成员组织所在国或地区服务时会引起知识产权拥有者与对方国家法律法规的冲突。作为外方的 UIA 成员组织提供设计服务时应充分考虑并提供合适的机会给东道主方的 UIA 成员组织，应通过提供公开的不保密的信息进行双方的合作。

8. 损害
辖区应当有权令侵犯版权的人支付合适的赔偿费给版权持有人。

9. 机构安排；最后条款
UIA 将监督本导则的实施，特别是监督 UIA 成员组织有义务遵守本导则。UIA 将向其成员组织提供与知识产权有关的咨询，UIA 将承担其成员组织委托的其他职责，特别是按照争端解决程序对有要求的成员组织提供协助。UIA 各成员组织同意在消除国际贸易中侵犯知识产权的原则下互相合作。

2001 年 11 月

附录

1. 建筑与职业制度

[美国]国际建筑师协会职业实践
委员会主任 J·席勒（James Scheeler）

章 岩 译

由于建筑工程的规模尺度和复杂性不断增加，传统建筑职业面临着许多挑战，国际设计合作将继续向前发展。

一些重要的问题：
- 新世纪对建筑工程的挑战和新的要求；
- 建筑服务范围和内容不断变化的趋势；
- 建筑师、业主、公众和社会相互之间的总体新关系；
- 建筑师和其他设计、建筑职业之间的新关系；
- 建筑师的资格标准和继续教育；
- 建筑师的职业义务和公共保护；
- 基于资质的选择和建筑师的委托；
- 建筑实践的新形式；
- 国际合作，建筑从业者的相互承认和相互的道德准则；
- 建筑职业制度和法律。

简介

今天，为了阐述"建筑与职业制度"这一问题，我将首先谈一谈关于职业和职业制度的一些一般性说法。然而，再讲一讲近来发生的正在推动建筑职业发展的事情，职业协会作出了什么反应，尤其是我们的国际建筑师协会作出了何种回答。

自从人类起源，我们一直生活在不断变化的环境之中，而且变化的速度正不断加快。我们中的一些人对此尤为敏感。人们为摆脱被时代淘汰的命运而付出的努力越来越多——我将我的感触归功于

"年事已高"!

简略地回顾一下历史和人类学，我们可以坚信，随着文明在岁月长河中不断进步，人类职业中越来越高层次的专业分工也不断演进并一直维系。专业分工引发了独特的知识技能体系的出现。随着专业分工的发展，我们需要推进崭新层次上的、社会各专业之间的理解、协调与合作。最终，这种进程演变成为我们所说的"职业"的发展。

《牛津英语辞典》中将"职业"定义为"一门工作，某些特定的知识或科学通过它应用于它所适用的其他事物上，或者通过它应用于以它为基础创立的一项工艺的实践上"。随后它提到三个学术职业：神学、法律和医学，并给出了一个更为宽广的定义——任何被个人惯常用来维持生计的行当或行业……

《牛津辞典》中将"职业人员"定义为："从事某种后天获得的，需要技能的职业的人"。随着社会的进步，其他专业群体步入我们所谓的"需要技能的"职业：会计师、建筑师、工程师等等。我想花一点时间来关注一下这一现象。

"职业制度理论"中，E·弗雷德森（Elioc Freidson）持有以下观点：一项职业是所有行业中的一种，而一个行业的一般行为表现为具体的工作。因此，职业随具体工作的变化而变化，它所需要的知识是定义一项职业的立法基础。他进一步将职业制度定义为对工作进行的行业化控制。他把两种情况进行对比，一种是由职业来组织和控制自己的工作，一种是更加普通的情况——由雇佣者（或劳动力的消费者）组织和控制工作，即：由雇佣者选择谁、什么时候去工作，并决定将做什么工作，怎么做。职业制度被用来代表对工作的行业化控制，这在逻辑和实践经验上都与消费者控制和经营者控制有着本质的区别。对工作的行业化控制是建立在高效率劳动力所需要的知识和技巧的基础之上的。

当国家组织起一个独立的知识与技能的实体并确立了劳动力或行业的分工，使之作为拥有那些技能的惟一团体时，便迈出了建立职业的第一步。这种行业化控制也建立了一个由加强准入要求所维持的庇护所：即只有那些获得证明其知识技能的职业证书的人可以被雇佣完成职业所限定的任务（比如做手术、教大学学生、在法庭上代表雇主、设计建筑和公共工程，或为正式财产声明作证）。通常，这些证书的确立包含了职业教育和培训的成型，这也是在制度上表明劳动力或职业分工就此划定。

职业培训的"工匠式"方法通常是在劳动力市场上的在职训练，由行业中被选作"教师"的人进行指导这一方法来源于"学徒"体系。另一方面，职业培训也通常在劳动力市场之外，在教室中，在实习的场所，由专职教员指导，他们持有该学科的证书，但不必参与职业实践。教师一般负责教学、研究和做学问。由于教师有教育机构的扶持，人们参与"纯粹"的研究从而扩展职业知识和技能是免费的。因此，职业中的这些成员可以拥有制度化的权威地位，而不必参与日常实践工作。他们传授标准的、被普遍接受的知识和技能，并可以制定职业成员用以判断的标准。这也常常造成学术工作者和实践者之间的紧张关系和不恰当的分野。

职业制度基础机构的主要调节变量是国家、职业协会、思想体系和不同职业的专门知识体系。这表明，职业制度机构在不能行使他们不具备的权力的情况下，无法成立或维持下去。在大多数情况下，只有国家拥有建立和维护职业和职业制度的权力。

建立这些职业制度的危险是传统的职业制度的标志——严格的准入，公开的、标准化的资格认证，以及道德行为准则的公开和政策化——可以被业主和公众视为是构建职业王国，推行排他，以及蓄谋与公众利益作对。

使职业制度发挥作用的主要手段，是使人们可以自由接触到专业知识，公开职业制度发展、应用和交流的进程，并在职业内外分享其知识。正是职业将不断前行的生活融入其知识体系。

如果你停下来去考虑自己国家的建筑职业，我猜测，你会承认它已经从主要依靠在职培训、以手工业为基础的行业，发展成为越来越依靠专职职业教育的一种职业。

对于设计行业，尤其是建筑师，教育的独特核心在于设计。对建筑师而言，设计是一个十分复杂的过程，包括许多其他的学科——历史、艺术、结构和系统工程、材料科技（这些学科的影响显而易见），然而社会科学、经济、商业管理、信息技术这些学科也影响到设计的决策，因而也必须成为建筑师知识基础的一部分。长期以来，建筑职业已形成了通过模拟工作室的方式使理论与实践相结合的独特教育方法——设计工作室。卡内基教学进步基金会的E·博耶（Ernest L.Boyer）博士和L·密特冈（Lee Mitgang）博士，他们通过在美国对于教育和实践的综合研究发现，在职业内部以及建筑与其他职业之间建立联系，是当今建筑职业所面临的惟一最为重要的挑战。值得庆幸的是，从建筑学最为独特的特征——设计工作室的天性和其传统而言，它具有将学问和知识整合、应用的巨大潜力。设计工作室是一个场所，同时也是一个过程：一种思考的方式，在这里，许多建筑知识的元素、可能性和限制被合为一体。

现在，我要考察一下国际建协对当代职业制度所面临的一些挑战的反应。

首先介绍一下背景。请记住，联合国预测2050年世界人口将达到93.8亿人。《国家地理》杂志1998年10月的人口专辑中曾鲜明地指出我们必须面对的一些问题。实际上，所有至2050年预计增加的人口将会出现在发展中国家。到那时，预计发展中国家的人口将从1998年的约计47.5亿增加到82亿。同期，发达国家的人口将从1998年的11.8亿下降到11.6亿。

最近一个时期，全世界范围内，从农村向城市中心的移民不断增加。

《工程新闻实录》1998年12月号上预测世界年度建设总水平为3,224.453亿美元。

我做了以国际建协各区域为单位的有关这一预测的计算与大家共同探讨：

请看以下计算。

它可以告诉你在你所处的区域中的建设投资水平。显而易见，这种在建设上的资源投资水平，加之我们城市地区内不断增加的人口，将对我们满足整个世界对食品和水的需求的能力，也对我们提供足够的社区设施和服务以满足不断扩展的社会的需求的能力产生巨大的影响。

国际建协的领导人员已经认识到在需要愈加复杂的机构和建筑的这个越来越复杂的国际社会，职业间和职业内合作的需要也越来越紧迫。建筑师的知识基础在不断变化并增长，若想维持其威信，还需要不断地维护和发展。然而，为国际建协制定职业规范和适宜角色是对我们的一个挑战。

作为以这一挑战的许多回答之一，国际建筑师协会理事会（UIA cuncil）于1994年6月成立了职业实践委员会并通过了它的纲领。美国建筑师学会和中国建筑学会组成了委员会的联合秘书处。通过职业实践委员会的纲领，国际建协寻求建立职业制度原则和符合公众健康、安全、福利以及市民的文化利益的推荐国际标准，并支持如下观点：职业制度标准和资格认定标准的相互承认是符合公众利益的，也是符合职业利益的。

委员会的早期工作项目是研究并记载全世界建筑实践的传统和实践标准。为得到数据，我向国际建协所有成员发放了调查问卷。经过多次试图得到重要回应的努力，我们还是仅收到了大约25%的回复。但这些已足以构架这一问题的范围。回复问卷的国际建协各成员国在建筑实践和程序上大不相同。显然，建筑这一职业以各种模式存在，极少有国际的标准，而且在一些国家里，这一职业尚未真正形成。在建筑作为一种职业的历史里，留下了许许多多社会经济和政治影响的烙印。英联邦建筑师协会拥有遍及全球的会员并在建立共同的教育标准方面进行了长期的努力。在《拿破仑法规》（Napoleonic Code）之后，许多国家的职业学校在学生完成规定课程后，同时颁发专业学位和专业职称。不同的是，有许多国家在授予职业证书之前要求一段时期的实习或职业培训并通过考试，而另一些国家则对建筑实践不作要求。

建立国际实践职业制度标准的兴趣是由建筑实践的不断全球化所引发的，也部分地受到《乌拉圭回合关贸总协定（GATT）》、欧盟的活动以及一些区域贸易协定的影响。毫无疑问，在过去的10年中，通讯条件的极大改善也是一个推动因素。

作为《乌拉圭回合关贸总协定》的结果，是建立起世界贸易组织。这一协定和与其相关的《服务贸易总协定（GATS）》（General

Agreement on Trade in Services)加速了世界向全球化社会、全球化经济、全球化建筑产业和全球化建筑职业迈进的趋势。

该协定所建立的"服务贸易委员会"负责制定有关资格认定的要求、程序和技术标准,以及授予执业证书的条件,该条件应在客观和透明的规范之基础上制定,而规范本身不能给服务贸易制造障碍。在判断是否履行此义务时,将考虑相关国际组织的标准。在适当的地方,认证将基于多边认可的准则和教育、经验、执照和证书发放的共同的国际标准。世界贸易组织负责为与联合国专门部门的协商和合作进行恰当的安排。

这些条款向国际建协及其成员提出了一个极大的挑战:带头迈向职业制度标准与建筑师资格认定标准(指向职业证书通用性)的相互承认。

从1994年到1996年,国际建协制定出一项限定职业制度的资格问题的政策文件(Policy paper)。这一政策文件题为"国际建协推荐的建筑实践职业制度国际标准的协定"。被国际建协理事会和1996年7月召开的巴塞罗那代表大会全票通过。

《协定》包括一项职业制度原则的说明和一系列政策问题。

职业制度的原则通过法律,也通过道德准则和限定职业行为的法规建立。

这些原则分四个方面:

- 专业能力,是指建筑师所拥有的确保其为公众恰当地提供专业建筑服务的,系统的知识体系、技能和理论;
- 严肃性,是指建筑师向业主提供专业服务而不受任何私利所左右;
- 承诺,是指建筑师以合格的、专业的方式为业主提供服务,执行公正的、毫无偏见的判断;
- 责任,是指建筑师应对其向业主提供的咨询意见负责。

"职业制度原则"之后,是以定义和背景说明形式阐述的一系列相关政策问题,其后是政策说明。

《协定》采用之后,国际建协大会把它的条款作为指导国际建协工作的政策。这一政策文件为职业实践委员会起草更为细致的推荐政策导则时的指导性文件，政策导则一经国际建协和其成员国通过，就可以成为职业的标准。1997—1999年的"职业实践纲领"致力于这些导则的制定和对《协定》第一版的适当修订。

　　《协定》和《政策导则》的目标，并非是寻求国际建协成员国间的既有标准和程序的协调。相反，它是通过国际建协成员国既有标准和实践与《协定》政策两相对比，发现差异或不同，并提出在多边基础上建立弥补差异的相应方法。《协定》的目的，不是在相互竞争的利益间建立由妥协性协议达成的强制性标准，相反，它是国际建筑师界相互合作，客观地建立最为符合社区利益的标准和实践的结果。《协定》和《政策导则》旨在定义什么是建筑职业中最为优秀的实践。它们将成为"活"的文件，会受到持续的检查，并在各方意见和实践经验表明必要的时候再进行修编。

　　国际建协致力于在职业的国际标准方面达成共识，毫无疑问，这将是一个漫长的过程；这一标准将是为达成相互认可协议而进行双边或多边协商的起点。

　　这一工作的进行，以下几点是必须坚持的：尊重每个国际建协成员国的主权，导则应允许应用替代原则的弹性，并在结构上为增补反映国际建协成员国地方条件的附加要求提供可能。

　　最初，这一努力的所有力量都用在使大家在"将政策文件的制订过程公开、透明化"这一问题上取得共识。国际建协的全体成员国都被邀请来提名委员会的代表。这一"公开委员会成员"的政策至今仍在实行。其结果，委员会的名单已增至57个国家的92名代表和10个国际、区域和民族组织的25名代表。委员会召开的会议向任何对委员会工作感兴趣的人敞开。

　　自1996年巴塞罗那《协定》通过以后，《协定》的条文在所有国际建协成员国间散发，并同时征集批评和建议。这些意见已在起草小组提议下，在《协定》的第二版条文中予以修改。

《协定》的政策问题

　　《协定》（第二版）涉及了被认为是建筑实践中职业制度的核心要素的16项政策问题：

- 建筑实践
- 建筑师
- 对建筑师的基本要求
- 教育
- 评估／确认／承认
- 实践经验／职业培训／实习
- 职业知识和能力的表达
- 注册／执照／证书
- 建筑工程设计任务的取得
- 道德和行为
- 继续职业发展
- 实践的范围
- 实践的形式
- 国外的实践
- 知识产权／版权
- 职业协会的角色

"政策问题"中1)建筑实践，2)建筑师，3)对建筑师的基本要求，这三方面的定义、背景和政策说明已经得到批准，在《协定》文件中已全部完成，正如现在所执行的那样。

建筑实践：按《协定》的语言，建筑实践包括提供与城镇规划设计、结构、扩建、保护、修复、建筑单体或群体的改建相关的职业服务。这些服务包括（但不限于）：城市规划和土地利用规划，城市设计，提供前期研究、设计方案、模型、图纸、说明书和技术文件，协调其他人员提出的技术文件（包括顾问工程师、城市规划师、景观建筑师和其他专业顾问……只要需要，没有限制），还有建设经济、合同管理、建筑监管（monitoring）［在一些国家称为"监理(supervision)"］以及工程管理。

建筑师："建筑师"的任命，一般是依照法律或惯例，授予能够在职业上和学术上胜任，并通常在他（或她）进行建筑实践的地区注册过的，持有执照的，或被授予证书实践的人。他们在空间、形式和历史文脉方面，对在该地区推广公平和可持续发展、福利以及社会人居环境的文化表达负有责任。

对建筑师的基本要求：对于建筑师注册、授予执照或证书的基本要求，包括须具备通过经评估的教育和培训所掌握的知识、技能和能力，以及可证明在职业上胜任建筑实践的知识、才能和经验。

1985年8月，欧洲社区委员会通过了一项《指令》，目前已成为在每个欧盟成员国执行的国家法律，它规定了建筑师所必须具备的知识、技能和能力。这项指令被推荐作为制定"国际建协国际标准"的最低基准。随着工作的进行和讨论在欧盟指令条款的基础上，一些其他内容加入其中。目前，已经有13个具体的标准：

· 创作同时满足美学和技术要求的，旨在达到环境可持续发展的建筑设计的能力；

· 足够的建筑和相关艺术的历史和理论，以及技术和人类科学的知识；

· 美术知识——建筑设计质量的一个影响因素；

· 足够的城市设计、城市规划知识和与规划过程相关的技能；

· 对于人与建筑，建筑与其环境之间关系的理解，对于联系建筑和建筑之间的空间以适应人类的尺度和需要的理解；

· 达到环境可持续设计的途径的足够知识；

· 理解建筑职业和建筑师在社会中的角色，尤其是在准备解释社会因素的说明书方面；

· 理解工程设计要点的调查和准备方法；

· 理解与建筑设计相关的结构设计、营建和工程问题；

· 足够的材料构造、技术以及建筑功能的知识，从而提供舒适的内部条件和对恶劣气候的抵御；

· 在成本和建筑法规条例所限制的范围内，满足建筑使用者需要的必要设计技能；

· 足够的将设计概念转变为建筑，把设计方案融入到整体规划之中所需要的产业、组织、规定和程序的知识；

· 足够的工程预算、工程管理和成本控制知识。

现在，以上三个政策问题已经制定完成，并在《协定》中占有一席之地。

对其余的政策问题，导则文件针对目前实践中出现的差异，特别予以均等原则，以前是这样；将来也是如此。

教育：建筑教育应该确保所有的毕业生具备建筑设计的知识和能力，包括技术系统和技术要求，以及对于健康、安全和生态平衡的考虑；应使建筑师理解建筑之文化的、精神的、历史的、社会的、经济的和环境的文脉；应使他们透彻地了解建筑师在社会中的角色和责任，这基于受到教育的、会分析的和有创造性的头脑。

依据《国际建协/联合国教科文组织关于建筑教育的宪章》，国际建协提倡，对于建筑师的教育不宜少于5年（不包括实践经验、职业培训或实习），应是经过评估、获得确认或得到承认的建筑学科的教育，同时应允许学校在教学方法上和对待地方文脉方式上的多样性，以及在达到同等效果的情况下的灵活性。

评估/确认/承认：《协定》的政策说明倡导：建筑教育课程必须是得到独立的相关权威机构评估、确认或承认的，这种机构脱离大学之外，并应以适当的时间间隔进行评估（通常不超过5年）。国际建协与相关国家高等教育组织共同协作，制定建筑师职业教育内容的标准，该标准是以学术为骨架，逻辑连贯，便于操作，注重结果，并以优秀实践所指导的措施来实行。

实践经验/职业培训/实习：《协定》的政策说明倡导：建筑学专业毕业生必须完成至少2年合格的实践职业培训或实习才能注册，得到执照或证书，从而成为建筑师（但要有工作到3年的标准），同时，也允许在达到同等效果的情况下有一定的灵活性。一项导则文件已经拟订出来，旨在协助国际建协成员国制定那些建筑系毕业生在注册、取得执照或证书开业之前，可接受的有指导、有计划的工作要求与程序。

职业知识和能力的表达：《协定》的政策说明倡导：建筑师的知识和能力必须以足够的证据来证实。这一证据必须包括在实践期、培训或实习结束时，至少通过一门考试。考试所不能涉及的必要的职业实践知识和能力，必须有其他适宜的证据。它们包括商业管理和相关的法律要求等内容。

一项协助国际建协成员国用来评价申报人在最初的教育和培训

后设计合用的建筑的综合职业实践知识的程序和考察办法的导则已经出台。

注册／颁发执照／证书：《协定》的政策说明倡导：国际建协在所有国家推行建筑师执业的注册、执照和证书发放的制度。为了公众利益，这种注册、发放执照和证书的条款应该具有法律效力。一项关于建筑师注册、执照和证书发放的法律认可的示范法规已经出台。

建筑工程设计任务的取得：《协定》的政策说明提倡：为了保证建成环境的生态可持续发展，并保护社会的文化、经济和社会价值，政府应该采用帮助选择最为适合该工程的建筑师的"取得程序"任命建筑师。这一提议只有在适宜的手段得到各方同意的条件下，才可能由以下方法实现：

• 建筑设计竞赛应该依据由联合国教科文组织和国际建协所确定的原则，并在国家政府或建筑职业协会同意下举办；

• 依据国际建协导则中制定的"基于资质的选择"（QBS）程序；

• 在完整的规定建筑服务质量、范围的说明书的基础上指导谈判。

一项名为"建筑师资质选择指南：通向质量的关键"的政策导则文件，已由国际建协和理事会制订、通过，并分发给国际建协各成员国。

道德和行为：一项道德和行为准则就可建立指导建筑师职业实践行为的"职业行为标准"，将1987年国际建协通过的《国际咨询服务道德准则》与《服务贸易总协定》的条款两相比较，可以看出，我们需要彻底地审视《国际咨询道德准则》。建筑师在这一问题上表达的许多忧虑与"建筑师在为一个社区设计建筑或建筑群时，应对其文化和价值体系给予适当的尊重和注意"的立法考虑相关。为了确保文化的完整性，国际建协的《准则》要求外国建筑师必须与当地建筑师合作。

《协定》的政策说明确认，目前国际建协的《国际咨询服务道德准则》仍然有效。国际建协鼓励各成员国将本国正在执行的道德和行为准则引入推荐的《协定导则》。只要不违背国际法或建筑师本国的法律，各成员国必须遵守建筑师所服务的国家或地区的道德行为准则。

继续职业发展：继续职业发展是一种维持、提高或增加建筑师的知识和持续提高能力的贯穿一生的过程。越来越多的职业团体和制定规章的机构要求他们的成员花时间以保持现有技能，扩展知识面，并探索新的领域。这一点，对于适应新技术、新实践方法以及不断变化的社会和生态条件而言愈加重要。为使其成员保持知识更新，职业组织可以要求其成员参加"继续职业发展"。

《协定》的政策说明倡导：为了公众的利益，国际建协强烈要求各成员国建立起"继续职业发展"制度，是作为其成员必须履行的一项义务。建筑师必须确信他们有能力提供他们所提供的服务，而且，行为准则必须强迫建筑师维护《建筑师的基本要求》所规定的各种地区里公认的标准（以及未来的变化）。同时，国际建协必须指导注册更新所需要的"继续职业发展"。在所有国家推行导则以帮助互惠关系的发展，并继续制定有关这一问题的政策。

一项简要的导则文件已经制定出来，涉及继续教育的问题和制订有关"职业发展要求的选择"的建议。

实践的范围：《协定》的政策说明倡导：国际建协鼓励和推动建筑实践范围的不断延伸（仅仅在道德行为准则条款的限制之下），并尽力确保范围延伸所需要的知识和技能的相应扩展。

实践的形式：《协定》的政策说明倡导：应该允许建筑师以其所服务的国家任何合法的形式进行实践，但总是要受制于主流的道德和行为要求。在国际建协认为必要的时候，将考虑实践的其他选择形式和不同的地方条件，制订并修改它的政策和标准。（如果这些形式是符合社会利益的，并能够扩展积极的、创造性的建筑职业角色。）

国外的实践：当建筑师个人或建筑公司，在他自己的国家之外寻求设计任务或已接受工程任务，或者提供相关服务时，我们称这是在国外的实践。大家普遍对加大建筑师可靠的流动性和提高在国外辖区内提供服务的能力表示兴趣。当然，我们也有必要加强对地方环境、社会和文化因素，以及对道德和法律标准的意识。

《协定》的政策说明倡导：在未经注册的国家中提供工程建筑服务的建筑师应该与地方建筑师合作，以确保法律、环境、社会、文化和遗产的因素能够得到适宜的和有效的理解。协作的限制条件应该仅仅由那些依据国际建协道德标准和地方法律法规所制定的条款所决定。

知识产权／版权：知识产权包含专利、版权、商标三个法律方面。它指的是设计者、发明者、作者和制造者的权利，包括他们的想法、设计、发明、作品以及产品和服务的来源认定。

尽管许多国家对建筑师的设计有一些法律保护，但这种保护常常是不够的。建筑师与新的业主讨论想法和概念后，业主却并不雇佣你，而又不费分文地使用建筑师的想法，这种情况并不鲜见。建筑师的知识产权在某种程度上被国际条例所保护。在《服务贸易总协定》的文本中，有对于与知识产权贸易相关问题的协定，包括侵权商品贸易（TRIPS），1955年9月16日的世界版权会议也具有国际意义。在欧洲，修订后的《1886年伯恩协议》在大多数国家仍然具有约束力。

《协定》的政策说明倡导：国际建协成员国的国家法律应该授权建筑师在对其权威和责任没有损害的前提下实践其职业，并保留他们作品的所有权和知识产权。

职业协会的角色：职业一般由制定标准（如：教育、道德规范和受监督的职业标准）的管理机构控制。规则和标准的设计旨在使公众获益，并不是照顾成员个人的利益。在有些国家，某些类型的工作是在法规的控制下保留给特定职业的，这不是为照顾协会的成员，而是出于保护公众利益的需要，因为这类工程必须由受过必要的教育和职业培训，遵守标准和纪律的人承担。协会的建立，旨在建筑的进步，知识的增进，并且通过确保其会员以公认的标准执业从而保护公众利益。

根据一个国家是否保护从业人员头衔和实际工作的不同（两者都保护或都不保护），职业协会的角色和责任有很大的不同。有些国家，立法机构也代表着职业，而在另一些国家，它们的职能是分离的。

职业协会的成员通常应该维护公认的标准。这是通过遵守由职业协会颁发的行为准则，履行成员的其他义务（比如"继续教育发展"）来实现的。

《协定》的政策说明提倡，在没有职业协会的国家，国际建协应该鼓励建筑从业人员依据公众利益成立这样的协会。

职业协会应该争取确保其成员遵守国际建协的国际标准，即《国际建协／联合国教科文组织关于建筑教育的宪章》的最低标准，以及《国际建协国际道德和行为准则》，并使其成员不断更新他们的知识和技能，以符合《基本要求》的规定（包括现在的以及未来的发展演进），并在总体上为建筑文化、知识的发展和他们所服务的社会作出贡献。

协定的政策导则：推荐的政策导则包括7个问题：
- 评估／确认／承认
- 实践经验／职业培训／实习
- 职业知识和能力的表达
- 注册／执照／证书
- 建筑工程设计任务的取得——基于资质选择
- 道德和行为
- 继续职业发展

国际建协职业实践委员会已成立了起草小组，拟订《实践范围》、《实践形式》、《外国的实践》、《知识产权／版权》、《职业协会的角色》等政策导则，将提交2002年在柏林召开的国际建协大会。

对于具体问题的安排：

将具体问题和任务分派给对该问题有真正兴趣的个人或团体的办法已被证实是非常有效的。委员会的秘书处已经邀请委员会成员和国际建协的成员国参加《协定》和《导则》的制定。已有足够多的具备《协定》中各个政策领域之丰富经验的国际建协成员国，为已提交委员会各会议讨论的工作文件的制订打下了坚实的基础。

对导则草案的评论：

随着准则草案的逐步制订，它们在成员间散发，以求得大家的评论和意见。这个过程会一直持续到准则正式提交国际建协理事会通过。经理事会通过之后，政策准则文件将会被出版，作为国际建协的补充政策文件和《协定》的一部分发给与会者、感兴趣的公众和政府部门。

《导则》一旦被通过，政策问题和相关的导则文件将散发至国际建协全体成员国，同时要求每个成员国将其推荐给政府和贸易官员，并通过他们，将其推荐至世界贸易组织。世贸组织的代表已参加了委员会的会议，熟悉了《协定》的内容。文件也将作为国际建协政策文件提交给世贸组织，以成为推荐的国际标准。国际建协希望通过这种方式，影响塑造国际服务贸易的政治进程。

但是，从总体上看，这些文件代表着建筑师国际社区所付出努力的第一阶段，他们要在为最好地维护社区利益而制定的标准和实践方面达成共识。《协定》和《政策导则》旨在定义当今建筑职业中什么是最优秀的实践。而这里我特别强调"当今"。因为它们是活的文件，应该接受职业的不断检验，并根据我们复杂世界中的经历和不可避免的变化进行修订。

结束本文之前，我想推荐《建筑学知识：职业的概念》一书，作者是英国皇家建筑师协会和欧洲建筑师委员会的前主席 F·达菲（Francis Duffy），并介绍他判断建筑职业制度的基础，以及判断建筑师职业协会（如国际建协）存在的依据。这一标准是：是否是自愿加入一个开放、独立、自我质疑、理性和艺术性的团体，是否有能力在建筑师之间、建筑师和公众之间进行持续的对话，致力于建立并分享关于建筑师所作所为和应作应为的知识，献身于发展建筑实践的最高可能标准。

我想强调的是，如果我们作为建筑师，不率先在我们最丰富的知识和经验的基础上建立最高的标准，这些标准将被政治家们和官僚们建立在他们的政治意愿上！

至少现在，我相信，我们的职业仍然还把握着它的全球命运。

2. 全球化时代的职业精神

[中国]中国建筑学会副理事长　张钦楠

摘要：本文试图探讨全球化对建筑设计实践的影响。首先研究全球化与扎根文化的冲突，尤其是在中国的情况；然后简要分析合作设计的利弊；最后讨论新文脉下的职业主义。

全球化来了！

经济学家断言，从20世纪60年代起，世界发生了根本变化，从在国界内组织本国经济转向在生产和贸易中实现综合和协调的全球性分工。

根据最近的数据报道，全球37000家跨国公司控制了世界生产和贸易总额的50%，海外投资的70%，新技术的80%，技术转让的90%。

这种全球化的趋势在中国大陆也很显著，从20世纪70年代末宣布"改革和对外开放的政策"以来：

——对外贸易从1978年占国家净产值的14.2%增至1994年的46.5%；

——实际直接利用外资从1979—1982年的12亿美元增至1994年的338亿美元。

——在建筑业中，使用外资的工程项目到1997年末达到39000项，合同投资额达1600亿美元。

——这种趋势也发生在建筑设计领域。"境外"（包括我国香港和台湾）来中国大陆注册承担个体项目设计的事务所、公司达数百家。许多中国城市景观的变化出自世界多国建筑师之手。

不论你喜欢与否，全球化已经降临，在21世纪更将如此，在中国是这样，在世界各地亦然。

现在的问题不是要不要全球化而是如何面临它，如何使它有利于我们，特别是有利于普通民众。

对建筑师职业来说，当职业实践跨越国界时，传统的职业精神不能简单地在地理上扩展，而是需要以更广阔的内涵和更新颖的视野来处理。

单向还是双向交通？

在建筑史中，建筑师的跨国设计由来已久。明显例子之一就是在北京的白塔寺，它是13世纪尼泊尔建筑师阿尼哥的作品。同样，在日本奈良，可看到8世纪中国建筑的影响。但只是在20世纪，这种跨国设计才有飞速的发展。

以20世纪中国为例，我们经历了3次高潮。第一次是世纪初（或可说在19世纪末就开始了），殖民主义的入侵带来了西方国家的建筑，如上海外滩。第二次发生在二战前后，现代主义的国际风格被输入中国，然而在50年代由"民族形式"的提倡所抵制。第三次是80年代改革开放后，世界各种建筑流派的思潮都涌入了中国。

回顾历史，我们可以说，第一次属于政治性的强制；第二次是意识形态的对峙；第三次是自愿的接受。这些输入，大量地属于"单向交通"，因其未考虑或不考虑本土扎根文化而受到抵抗。

当然，这样的论断过于简单化。历史证明，外来影响，不论是强加的还是自愿接受的，在文化交融中往往产生出乎意料的影响。例如，在第一次殖民主义的侵入中，中国一些沿海城市中出现了一种新的居住模式：如上海的里弄建筑，它可以说是欧洲工业革命后在城市中开发的联排工人住宅与中国江南地区联排四合院的"跨文化"产物，它们很适应当时中国的国情。在第二次输入中，尽管国际风格在1950年代在理论上被否定，但中国的现实经济条件却产生了大量砖混结构的方盒子建筑，可称之为"没有格罗皮乌斯的包豪斯"。这也是一种"跨文化"的产物，也很适应中国的国情。第三次输入的时间尚短，很难说是否又产生了新的"跨文化"产品，但是贝聿铭先生的香山饭店和吴良镛教授的菊儿胡同，却显然具有外来和本土文化交融的特征。

我首次接触"跨文化"建筑一词是在阅读K·弗兰姆普敦教授

对 J·伍重的创作道路的分析。他描述了伍重如何在自己的许多设计中融合了西方、中国、日本的文化观念和表达手法，体现在悉尼歌剧院、巴格斯瓦德教堂等设计中，并且在科威特国民议会大厦中又揉合了阿拉伯的文化特征，堪称为自觉运用"跨文化"手法的建筑师的杰出代表。

在20世纪末的今天，当我们回顾历史时，殖民主义的古典主义经念碑已成为近代文物，现代主义的国际风格往往成为忽视地方条件的批判对象，它们的成败告诉我们一个真理；不论你意识与否，当不同文化对撞时，最有生命力的往往是自觉或不自觉地产生的非此非彼、亦此亦彼的"跨文化"产物。到了20世纪末，我们可以看到更多的建筑师认同"地方精灵"、文化认同、乡土气息、文脉主义。许多建筑师在跨越国界时，比在本国设计更为谨慎，更多注意当地的地理、气候和文化特征。可以说，"单向交通"正在逐步让位于自觉的"双向交通"，在全球化的趋势中，这种转变是值得欢迎的，并且势必成为21世纪建筑师职业精神的一种基本观念。

合作设计是否违反市场道德？

1987年，UIA在杜布林代表大会上基本采纳了此前由亚洲建协拟订的《国际咨询服务的道德准则》。该准则的核心思想是要求外国设计师尊重项目所在国的文化，并与所在国的建筑师合作设计。近年来，特别是世界贸易组织的GATS总协议出台之后，对这个准则是否在精神和文字上符合GATS要求，以及合作设计是否符合自由贸易的市场准则和道德精神是存在争论的。

仍以中国为例。在我们对外开放建筑设计市场时，中央和地方的有关规定都要求境外（包括港澳台）建筑师必须与中国建筑师合作设计，这已成为习惯做法。

人们可以说，所以如此规定，是由于中国的市场开放还处于早期状态，中国的注册建筑师制度正在建立，有关设计合同、设计义务、争端仲裁等尚待完善。与此同时，许多境外建筑师对中国国情还很生疏，在这种条件下，合作设计有其存在的逻辑性。

然而，经过10余年的实践，我们认为，除了以上这些客观因素外，合作设计本身具有一系列的优越性，如：

• 帮助境外建筑师更好地了解当地条件（气候、地质、建材、施工、规范、生活方式、风俗习惯等）；

• 由于当地建筑师更了解本国的建筑法规和程序，可使设计意图更好地被有关当局和部门所接受；

• 由于利用地方人力资源，使设计成本能够降低；

• 与承包商的联系更为直接，解决施工问题更为迅速；

• 地方当局由于有本国建筑师参与设计而更为放心；

• 从长远来说，合作设计促使境内外建筑师更为相互了解，推动了建筑师职业素质的全面提高。

诚然，合作设计并不是"十全十美"的，特别是合作对象（内外）水平不高、合作诚意不足，或有地方保护主义的干扰等因素存在时会影响甚至取消合作效果。但总的说来，在建筑师资格的互相承认还只局限于少数国家和地区时，合作设计是当今条件下最有效的一种跨国设计的实施方式。在全球市场还处于分割状态下，它可以促进开放、服务贸易和职业流动性。

全球化过程中的职业精神

为了保护和发扬高尚的职业精神，各国许多职业团体根据多年的实践经验制订了职业道德准则，但这些准则一般都只适应于国界以内。显然，全球化的发展将使各职业界（包括建筑师职业）面临新的前景，即全球性执业的前景。

尽管一些基本原则仍然有效，在全球化形势下的职业精神将不会是原有的限于国界内的职业精神的简单扩大，一旦人们跨越了国界，就会产生新的内容和新的原则。UIA职业实践委员会从1994年开始就研究如何面对这一新的现实，并在许多成员组织的积极参与下取得了良好的成果。

据我看来，这些新的内容和原则包括：

• 遵守所执业国家的法律和规定，同时又不违反本国的法律和规定；

- 关注所执业国家的环境、地方条件和文化，在尊重业主和地方当局有关要求的同时，要关注当地公众的利益，保护当地的环境和文化；
- 在与当地建筑师合作设计时，把设计质量和效益放在首位；
- 培育和当地建筑师与公众的相互了解等。

我认为，在跨国设计中存在的一个最重要的问题是如何处理好全球化与扎根文化的辩证关系，使各种民族的、地区的和地方的文化能继续为全球文明作出贡献，而全球文明又反过来推动民族、地区和地方文化的更新和发展。

在20世纪的历史中，殖民主义建筑的一个动机就是树立殖民者的权威，它的出发点就是否定和贬低所征服民族的原有文化。现代主义的国际风格与殖民主义显然不同，但是它的哲学观点中也往往过分强调了科学技术的全球性，而忽视了民族和地方特色。在我看来，20世纪建筑学的最杰出的成就，就是许多建筑师在自己的作品中证明了全球性文明和民族、地区文化的相互可容性。

当然，表现这种可容性的手法和程度是多种多样的。人们常常说，现在的世界是多元化或多极化的，建筑表现也必然是多元化的。在同意这一客观描述的同时，我们不能不注意到，在文化史中，多元化和折衷主义一样，往往只是个过渡现象。

在自然界，元素和元素往往首先以混合物形式共存，然后自然地（如氢氧成水）或人为地（如合金和有机塑料）变为稳定的化合物。建筑也是如此，殖民主义建筑企图否定地方文化，但是却不自觉地产生了像上海里弄那样的化合物，地方文化也企图拒绝国际风格，也不自觉地产生了像中国20世纪60—70年代大量建造的筒子楼。与此同时，越来越多的建筑师正在自觉地创造各种跨文化的建筑物。在我看来，多元化的意义不在混合，而在不断地化合。

21世纪世界建筑师的职业精神，就是在接受全球性的同时，承认各民族、地区和地方文化的价值，以平等合作的方式，创造出丰富多彩的，跨文化的新建筑。

3. 较自由的服务贸易市场

——2002年7月24日在柏林第21届
世界建筑师大会上的演讲

WTO律师,《国内法规工作组》秘书　D·B霍纳克

下午好,谢谢国际建筑师协会邀请我作演讲。首先介绍一下我自己,我在WTO的服务贸易局工作,有十年时间从事WTO/GATT工作,第一个五年在贸易政策评估局工作,工作内容是从经济角度检验成员国的贸易政策,后五年在服务贸易局工作,我所主要负责的领域是国内法规、专业服务和旅游业。

我演讲的题目是"较自由的服务贸易市场"而不是"自由的服务贸易市场",这可能会使你们中的某些人感到意外,因为至少有一位记者将WTO描述为"自由贸易的狂热者",如果你仔细地看一看《服务贸易总协定》(即GATS),你就可看到协定中没有自由贸易的意思,然而确能找到与渐进式自由化有关的内容,我想提醒注意的是,即使在货物贸易领域,我们已有五十多年的贸易谈判,我们仍然没有自由贸易,因此对于服务贸易领域,包括建筑师的服务贸易,达到自由贸易是不现实的。

今天讲与国际建筑师协会《认同书》和WTO贸易自由化努力有关的四方面内容:

(1) UIA制定的是推荐标准,而不是最低标准,目的是保证建筑师服务的一定质量,这是不是UIA的工作和WTO的打算不一致?

(2) WTO的下一步做什么?WTO现正在做什么?WTO国内法规局将做什么?

(3) 如果UIA认同书继续发展的话应涉及其他什么问题?

(4) WTO和建筑师职业的合作如何增强?

我将演讲分为四个部分,先讲服务谈判,作一个概括介绍,即服务谈判如何进行及服务谈判的内容。然后谈谈有关"国内法规"，以 WTO 术语来说，大部分的《UIA 认同书》的条款属于"国内法规"，最后，我想专门谈谈《服务贸易总协定》(GATS)和国际建筑师协会的《认同书》。

服务谈判

WTO 目前正在进行广泛领域内的贸易谈判。服务谈判实际上于 2000 年开始，这是按照《服务贸易总协定》(GATS)第十九条进行的，你们会注意到 GATS 的实际条文是多么短，人们常常听到的所谓的几百页实际上是 WTO 成员政府的承诺表。这些承诺表可能对于一个发达国家会超过 100 页，而对于最不发达的国家只有 1 页。

根据 GATS 总协定的第十九条，要求在渐进地自由化(progressive liberalization)基础上进一步谈判，按照 WTO 各成员国所制定的指引和程序，每一阶段的谈判都设立了最后期限，2002 年 7 月 30 日是成员国送给贸易伙伴关于进一步将 GATS 自由化要求的最初期限，第二阶段是从 2003 年 3 月 31 日开始，当 WTO 成员们在收到所有的要求之后，将开始提出他们对于服务自由化的许诺，总的谈判截止期限是 2005 年 1 月 1 日。

重要的是区分"要求"和"许诺"，"要求"通常是对某一个成员国提的，根据 GATS 最优惠国（MFN）条款，许诺应无歧视地准对所有 WTO 成员国。这一点非常重要，因为大国之间的谈判会涉及所有的国家。同样如果大国未涉及有关的内容就只能将中小国感兴趣问题提出来。由于发展中国家数量占了 WTO 成员的 80%，任何发展中国家想提交的问题，如果他们协同起来，都会被重视。

在最初阶段，WTO 成员已开始传发谈判提案，提升了什么是主要谈判目标，分清最大的贸易障碍。关于建筑师的服务贸易，到目前为止只有一项专门的来自澳大利亚的提案。我想指出所有的谈判提案及其他许多信息都可在 WTO 网站(www.wto.org)找到。澳大利亚的提案分三部分，第一部分是关于建筑师服务贸易自由化的重要性，第二部分列出了建筑师服务贸易自由化的阻力，第三部分是澳大利亚提案的建议。

该提案中所列出的贸易阻力是：限制设立建筑师机构，限制汇出利润；税收歧视；过度繁琐和不透明的签证程度；当地公民或居民才可执业；限制或不认可外国资质；强制性的职业团体会员要求；不透明的法规环境。

澳大利亚的提案要求消除这些障碍，努力促进建筑师资格互认，有效促进建筑师垮境流动等。除了澳大利亚的提案，还有7个其他WTO成员已提交了谈判的提案，建议将专业服务作为一个整体来谈判，这些国家和地区是加拿大、智利、哥伦比亚、欧盟、日本、肯尼亚和瑞士。挪威在他们的商业服务提案中也包括了专业服务，像澳大利亚提案一样，这7个提案也列出了贸易壁垒及对于进一步自由化的建议。

现在已有120个谈判提案提交，发达国家的提案和发展中国家提案接近对等。然而你如果考虑WTO中有80%成员是发展中国家，那么实际上是不平衡的，这些120个提案几乎包括了所有的服务领域，据我所知，健康服务领域是惟一没有被提案涉及的，我们现已基本结束提交提案的阶段，下面将进入"要求"阶段，各国正忙于对他们的贸易伙伴提要求并处理这些进一步自由化的要求。

如果我们看一看现在的GATS承诺的数目，包括乌拉圭回合及后来的谈判，以及WTO新成员的承诺，我们可以看到建筑师的服务已达到中等水平的成功。依照WTO的标准，这个水平是高的。现已有58个WTO成员(包括欧盟作为一个整体)对建筑师服务作了承诺。

需要注意的是，GATS总协定的灵活性允许成员国的实际的市场自由化程度有不同的时间表。对GATS作出承诺并不意味着WTO成员有义务实现有关领域的完全自由化贸易。事实上，协定中的灵活性允许成员在某一领域很大程度开放或很少对贸易伙伴开放。因此，重要的是，成员国所作的承诺并不代表这些承诺的质量。

国内法规

服务贸易谈判导则和程序指出，成员国应按GATS第6条4段涉及国内法规的前提下完成谈判。关于国内法规的谈判应优先于专门承诺谈判的成果，也就是说，要在市场准入和国家条款的决策之前先完成国内法规建设问题。

给你们一些背景情况,在乌拉圭回合谈判中,各国政府认识到,一些非歧视性的国内法规,如消费者保护和质量要求等等,也会有潜在的贸易限制效果。各成员国同意需要制定专门的规定来保证国内法规措施不搞烦琐或过分的贸易限制,我强调"过分"是因为理解到现实中这些国内法规有时需要些烦琐或贸易限制。WTO目标是将这类限制和烦琐内容减到最小程度。其结果就是 GATS 规定,即执业资格证书要求和程序,认证要求和程序及技术标准。在该条款中规定对任何专业人士必须制定这三个方面的相关内容。

大家已认识到各国政府对于专业服务的法规很多,因此在实施第 6 条第 4 段时,在 1995 年的讨论专业服务时各政府贸易部长有一个决定:设立专业服务工作组。我以前就在那个工作组担任秘书,经过几年的讨论,包括有关国内法规中的市场准入和国家条约规定,WTO 成员们于 1998 年 12 月成功地制定和采纳了《会计工作领域国内法规要则》,会计师工作专业要求的条例是很具有原则性的,它可以容易运用到其他专业,如建筑师工作,也可能对大多数专业服务都适用。该规定只有 4 页 26 个段落,分为 8 个部分:目标、总的条款,透明性(5 条)、执照要求(6 条)、执照程序(5 条)、认证要求(3 条)、认证程序(3 条)、技术标准(2 条),会计工作条例还没有在法律上实施,WTO 成员国决定等到目前这一轮谈判有成果后将市场准入和最惠国待遇作为一个整体,提交所在国的立法机构批准。有一点需要强调的是,即使一个 WTO 成员国在日内瓦签了 WTO 协定,该协定需要该国的议会批准才在法律上生效。另外要说明的是《会计师工作专业要则》。1999 年 4 月,服务贸易委员会(它是 WTO 服务谈判和总的服务执行机构)开始制定以 GATS 第 6.4 条为指导的法规性专业要求,该委员会设立了一个国内法规工作组,我在该工作组任秘书。该工作组有一位主席,他是一国政府的代表。新的国内法规工作组工作内容包括了以前专业服务工作组的工作。但不仅仅是专业服务,而是准备所有服务都适用的有关要求,该工作组也制定各服务领域包括专业服务领域的职业资格要求。我们有两方面工作,一方面我们称为方向的要求,如保证国内透明，这方面的要求对于所有服务领域可能都适用，另一方面是专业服务的工作。

国内法规工作组最初的重点是有关专业资格概念的讨论，有四方面：总则、透明度、等效性和国际标准。

目前工作组的重点在检验实际法规，WTO成员已送交了他们认为可以将其他职业资格要求都包括的法规条款，工作组现在阅读这些已提交的文件，目标是普遍适用的职业资格的相关组成部分，职业资格是强制性的，而导则，如1997年WTO所制定的《会计工作互认协议和安排导则》是非强制性的。WTO成员在他们制定国内专业服务中继续参与和探讨会计工作条例的普遍适用性。《服务贸易总协定》(GATS)和《国际建筑师协会认同书》。最后，我要讲的是GATS和UIA认同书的关系问题，我还有三个问题要回答。第一个问题是："如果我们同意推荐标准而不是最低标准来保证建筑师服务工作的一定质量，这是否与WTO的打算一致？"回答是很肯定的。

GATS中第6.4条所包括的三种规则中，很遗憾的是技术标准受到的关注最少。然而技术标准最终能证明是国内法规中最重要的内容之一，关于这一点，应当说明的是WTO不是一个制定标准的机构，GATS协定并没有告诉WTO成员必须有什么水平的法规，或者这些法规中应包括哪些内容，GATS的目标是保证这些法规不要过于烦琐或过分严格即限制了贸易，用国际标准是达到这个目标最好的方法之一。

GATS第7条是关于互认的，实际上在第4段中规定了WTO成员应与政府和非政府组织合作，以设立相关贸易或专业实践的国际标准，参加制定国际标准是WTO成员国必须做的，有意思的是WTO关于服务的国际标准，如在GATS中所包含的那样，要比货物贸易中的弱，在货物贸易中，基本上要求WTO成员采纳国际标准，除非他们有很好的证明为什么不采纳的理由。

第二个要回答的问题是，《UIA认同书》应继续在其他哪些问题上工作，我个人的看法是，《认同书》已经强调了所有主要问题，我对于UIA认同书的全面性印象很深。但我还有些建议，首先是可看一看澳大利亚谈判提案所列出贸易阻力，也可看看其他服务贸易的提案，已制定和通过的国际标准可以极大地有助于建筑师工作互认协议，帮助了跨境认证。这两大问题是目前服务谈判中的主要问题。其他方面的进一步工作可能包括了改进与建筑师服务有关的国内法规，如国际透明度，减少对居住地要求的贸易限制，也许最重要的工作之一是帮助发展中国家全部使用《UIA 认同书》。

第三个问题是：如何增加 WTO 和专业之间的合作？我还是从个人的观点来谈，有几个建议，上周 UIA 在 WTO 举办了一个研讨会，介绍了他们在国际标准方面的工作，特别是关于《UIA 认同书》，这类研讨会以及 UIA 最近的工作，与 WTO 和国内法规工作组的日常工作的相结合的意义是难以估量的。另外，UIA 及其各国成员可以起到与建筑师职业有关的技术信息源的作用，大家要记住，大部分外交人员不是专家，也不是建筑师。UIA 各成员国组织可以对于《会计条例》提出建议，因为 WTO 正在对该条例征询意见，看是否可以推广到其他的专业服务。同时鼓励 UIA 的各成员国组织与他们自己国家的政府沟通，支持目前服务谈判中贸易自由化努力。

对于 WTO 工作的体会（回答关于将《UIA 认同书》在 WTO 成员国中传阅的问题）

我代表 WTO 秘书处，但只有 WTO 成员国政府代表才能作出实际决策。作为秘书，我能说的是这些 WTO 成员的代表们已经同意的问题。

现实是 UIA 成员必须与 WTO 政府成员一道工作，WTO 是一个成员决策组织，你们不能简单地与秘书处一起工作，这里没有捷径程序。对 UIA 来说，作为一个非政府组织与秘书处沟通和提供信息是有用的，秘书处准备了背景材料，并提供了建议给 WTO 成员国和主席。但是，秘书处无权颁发《UIA 认同书》。

4. 多哈回合谈判澳大利亚政府关于建筑服务贸易的提案

世界贸易组织

S/CSS/W/63

2001年3月28日

(01 – 1526)

服务贸易委员会

澳大利亚关于建筑服务贸易谈判提案

下列澳大利亚代表团的意见已收到，按来件要求在服务贸易委员会的成员中通告。

本文是澳大利亚关于建筑学服务领域的谈判提案。澳大利亚保留以后进一步深化的权利。

1. 建筑服务贸易自由化的重要性

(1) 建筑服务贸易是所有世贸组织成员经济基础的重要组成部分，如WTO(S/C/W/44)条款所提出的，建筑学（以及工程的）服务的产值的经济重要性不是很直接就能算出来，在官方统计中，通常被统计在商业服务或建筑活动。但毫无疑问，建筑服务在多数WTO成员国是重要的专业服务领域，在许多国家，建筑和工程服务的出口所获得的境外收入比建造及其工程服务领域要多，建筑服务贸易可能主要由小型企业提供，其主要成本是高水平员工的薪水。

(2) 建筑服务已在国际间形成的贸易，主要通过在东道国立外国商务机构，通常由有专业技能的员工的临时性流动来提供服务。

(3) 通过服务贸易总协定谈判推动建筑服务贸易进一步自由化，预期能在更大更深的范围的地区市场提供水平更高的建筑服务，转移技术和技能，降低成本。

2. 进一步自由化的障碍

澳大利亚的建筑服务出口商列出了建筑服务进一步自由化的障碍：

- 限制设立商业机构
- 限制汇出利润
- 税收中的歧视
- 签证程序的烦琐和不透明
- 对于居住地和公民身份的要求
- 限制或不承认国外的资格
- 强制性要求职业团体成员
- 不透明的法规环境

3. 提案

澳大利亚的提议是：

成员国以消除壁垒的眼光来审视以任何形式限制设立商业机构是不公平的。

对于外国建筑师注册执业资格的要求和程序应在必要的测试范围内，如澳大利亚 2000 年 9 月 15 日的文件（S/WPDR/W/8）所述，它要求任何国内法规应符合对贸易最小限制目标。

成员国应全力实施服务贸易总协定第 7 条第 3 和 5 段，并尽最大努力使达成建筑服务资质的互认。

成员国核查与自然人流动的相关规则，以保证已获批准在另一成员国执业的建筑师能有效和迅速地获得签证。

建筑师服务的职业要求和会计师职业的要求一样（见 WTO 文件 S/L/64），应加强和调整使它们能满足建筑服务出口的需要。

译者注：本文所提到的建筑服务均指 Architectural Services，即建筑师及其相关人员提供的服务。

International Union of Architects

UIA Accord on Recommended International Standards of Professionalism in Architectural Practice

Third Edition
Adopted by the XXI UIA Assembly
Beijing, China, June 28, 1999
Preamble Adopted by XXII UIA Assembly
Berlin, Germany, July 27, 2002

UIA Professional Practice Program Joint Secretariat

The American Institute of Architects
Co-Director James A.Scheeler, FAIA
1735 New York Avenue, NW
Washington, DC 20006
Telephone: 202 – 6267315
Facsimile: 202 – 6267421

The Architectural Society of China
Co-Director Zhang Qinnan, Vice President
Bai Wan Zhuang, West District
Beijing, China 100835
Telephone: 8610 – 88082239
Facsimile: 8610 – 88082222

Contents

Preamble	98
Introduction	99
UIA Accord on Recommended International Standards of Professionalism in Architectural Practice	101
Principles of Professionalism	101
Policy Issues	103
Practice of Architecture	103
Architect	103
Fundamental Requirements of an Architect	104
Education	106
Accreditation/Validation/Recognition	106
Practical Experience/Training/Internship	107
Demonstration of Professional Knowledge and Ability	107
Registration/Licensing/Certification	108
Procurement	109
Ethics and Conduct	110
Continuing Professional Development	110
Scope of Practice	111
Form of Practice	111
Practice in a Host Nation	112
Intellectual Property and Copyright	113
Role of Professional Institutes of Architects	113
Appendix A	115

Note: Guideline Documents have been prepared and approved for the following Policy Issues of the Accord:

    Accreditation/Validation/Recognition
    Practical Experience/Training/Internship
    Demonstration of Professional Knowledge and Ability

Registration/Licensing/Certification
Procurement - Qualification Based Selection
Ethics and Conduct
Continuing Professional Development
Practice in a Host Nation
Intellectual Property and Copyright

Preamble

As professionals, architects have a primary duty of care to the communities they serve. This duty prevails over their personal interest and the interests of their clients.

In a world where trade in professional services is rapidly increasing and architects are regularly serving communities other than their own, the International Union of Architects believes that there is a need for International Standards of Professionalism in Architectural Practice. Architects who meet the standards defined in this Accord will, by virtue of their education, competence and ethical behavior, be capable of protecting the best interests of the communities they serve.

Introduction

The UIA Council established the Professional Practice Commission and approved its program in 1994. Following some 25 months of intensive activity by the Commission during the 1993-1996 triennium, the UIA Assembly unanimously adopted the first edition of the Proposed UIA Accord on Recommended International Standards of Professionalism in Architectural Practice in Barcelona, Spain in July 1996. By this action of the UIA Assembly, the Accord was established as policy recommendations to guide the ongoing work of the UIA and the UIA Professional Practice Commission.

The first edition of the Accord was transmitted to all member sections of the UIA with the request for their comments and cooperation in the further development of the policy framework for presentation to the XXI UIA Assembly in Beijing, China, in 1999. The 1997-1999 Professional Practice program focused on responding to comments and recommendations received from Council members, UIA member sections, and members of the Commission on the Accord and its policies. The first edition of the Accord was modified in response to those comments and as a result of Commission debate of the policy issue guideline documents being developed to flesh out the bare bones policy framework of the Accord.

The Accord and guidelines recognize the sovereignty of each UIA member section, allow flexibility for principles of equivalency, and are structured to allow for the addition of requirements reflecting local conditions of a UIA member section.

It is not the intention of the Accord to establish obligatory standards set by negotiated agreements between competing interests. Rather, the Accord is the result of the co-operative endeavor of the international community of architects to objectively establish standards and practices that will best serve community interests. The Accord and Guideline documents are intended to define what is considered best practice for the architectural profession and the standards to which the profession aspires. These are living documents and will be subject to ongoing review and modification as the weight of opinion and experience dictates. Whilst respecting the sovereignty of UIA member sections, they are invited and encouraged to promote the adoption of the Accord and the Guidelines and, if appropriate, seek the modification of existing customs and laws.

It is intended that the Accord and guidelines will provide practical guidance for governments, negotiating entities, or other entities entering mutual recognition negotiations on architectural services. The Accord and guidelines will make it easier for parties to negotiate recognition agreements. The most common way to achieve recognition has been through bilateral agreements, recognized as permissible under Article VII of the GATS. There are differences in education and examination standards, experience requirements, regulatory influence etc., all of which make implementing recognition on a multilateral basis extremely difficult. Bilateral negotiations will facilitate focus on key issues relating to two specific environments. However, once achieved, bilateral reciprocal agreements should lead to others, which will ultimately extend mutual recognition more broadly.

The Accord begins with a statement of "Principles of Professionalism," followed by a series of policy issues. Each policy issue opens with a definition of the subject policy, followed by a statement of background and the policy.

The XXI UIA Assembly in Beijing, China unanimously adopted the Accord in June 1999. A copy of the Resolution of Adoption is attached as Appendix A.

UIA Accord on Recommended International Standards of Professionalism in Architectural Practice

Principles of Professionalism

Members of the architectural profession are dedicated to standards of professionalism, integrity, and competence, and thereby bring to society unique skills and aptitudes essential to the sustainable development of the built environment and the welfare of their societies and cultures. Principles of professionalism are established in legislation, as well as in codes of ethics and regulations defining professional conduct:

Expertise: Architects possess a systematic body of knowledge, skills, and theory developed through education, graduate and post-graduate training, and experience. The process of architectural education, training, and examination is structured to assure the public that when an architect is engaged to perform professional services, that architect has met acceptable standards enabling proper performance of those services. Furthermore, members of most professional societies of architects and indeed, the UIA, are charged to maintain and advance their knowledge of the art and science of architecture, to respect the body of architectural accomplishment, and to contribute to its growth.

Autonomy: Architects provide objective expert advice to the client and/or the users. Architects are charged to uphold the ideal that learned and uncompromised professional judgment should take precedence over any other motive in the pursuit of the art and science of architecture.

Architects are also charged to embrace the spirit and letter of the laws governing their professional affairs and to thoughtfully consider the social and environmental impact of their professional activities.

Commitment: Architects bring a high level of selfless dedication to the work done on behalf of their clients and society. Members of the profession are charged to serve their clients in a competent and professional manner and to exercise unprejudiced and unbiased judgment on their behalf.

Accountability: Architects are aware of their responsibility for the independent and, if necessary, critical advice provided to their clients and for the effects of their work on society and the environment. Architects undertake to perform professional services only when they, together with those whom they may engage as consultants, are qualified by education, training, and/or experience in the specific technical areas involved.

The UIA, through the programs of its national sections and the Professional Practice Commission, seeks to establish principles of professionalism and professional standards in the interest of public health, safety, welfare, and culture, and supports the position that inter-recognition of standards of professionalism and competence is in the public interest as well as in the interest of maintaining the credibility of the profession.

The principles and standards of the UIA are aimed at the thorough education and practical training of architects so that they are able to fulfill their fundamental professional requirements. These standards recognize different national educational traditions and, therefore, allow for factors of equivalency.

Policy Issues

Practice of Architecture

Definition:

The practice of architecture consists of the provision of professional services in connection with town planning and the design, construction, enlargement, conservation, restoration, or alteration of a building or group of buildings. These professional services include, but are not limited to, planning and land – use planning, urban design, provision of preliminary studies, designs, models, drawings, specifications and technical documentation, coordination of technical documentation prepared by others (consulting engineers, urban planners, landscape architects and other specialist consultants) as appropriate and without limitation, construction economics, contract administration, monitoring of construction (referred to as "supervision" in some countries), and project management.

Background:

Architects have been practicing their art and science since antiquity. The profession as we know it today has undergone extensive growth and change. The profile of architects' work has become more demanding, clients' requirements and technological advances have become more complex, and social and ecological imperatives have grown more pressing. These changes have spawned changes in services and collaboration among the many parties involved in the design and construction process.

Policy:

That the practice of architecture as defined above be adopted for use in the development of UIA International Standards.

Architect

Definition:

The designation "architect" is generally reserved by law or custom to a person who is professionally and academically qualified and generally registered/licensed/ certified to practice architecture in the jurisdiction in which he or she practices and is responsible for advocating the fair and sustainable development, welfare, and the cultural expression of society's habitat in terms of space, forms, and historical context.

Background:

Architects are part of the public and private sectors involved in a larger property development, building, and construction economic sector peopled by those commissioning, conserving, designing, building, furnishing, financing, regulating, and operating our built environment to meet the needs of society. Architects work in a variety of situations and organizational structures. For example, they may work on their own or as members of private or public offices.

Policy:

That the UIA adopt the definition of an "architect" as stated above for use in developing UIA International Standards.

Fundamental Requirements of an Architect

Definition:

The fundamental requirements for registration/licensing/certification as an architect as defined above, are the knowledge, skills, and abilities listed below that must be mastered through recognized education and training, and demonstrable knowledge, capability, and experience in order to be considered professionally qualified to practice architecture.

Background:

In August 1985, for the first time, a group of countries came together to set down the fundamental knowledge and abilities of an architect (*). These include:

- Ability to create architectural designs that satisfy both aesthetic and technical requirements, and which aim to be environmentally sustainable;
- Adequate knowledge of the history and theories of architecture and related arts, technologies, and human sciences;
- Knowledge of the fine arts as an influence on the quality of architectural

design;
- Adequate knowledge of urban design, planning, and the skills involved in the planning process;
- Understanding of the relationship between people and buildings and between buildings and their environments, and of the need to relate buildings and the spaces between them to human needs and scale;
- An adequate knowledge of the means of achieving environmentally sustainable design;
- Understanding of the profession of architecture and the role of architects in society, in particular in preparing briefs that account for social factors;
- Understanding of the methods of investigation and preparation of the brief for a design project;
- Understanding of the structural design, construction, and engineering problems associated with building design;
- Adequate knowledge of physical problems and technologies and of the function of buildings so as to provide them with internal conditions of comfort and protection against climate;
- Necessary design skills to meet building users' requirements within the constraints imposed by cost factors and building regulations;
- Adequate knowledge of the industries, organizations, regulations, and procedures involved in translating design concepts into buildings and integrating plans into overall planning;
- Adequate knowledge of project financing, project management, and cost control.

Policy:

That the UIA adopt a statement of fundamental requirements as set out above as the minimum basis for development of UIA International Standards and seek to ensure that these particular requirements are given adequate emphasis in the architectural curriculum. The UIA will also seek to ensure that the fundamental requirements will be constantly kept under review so that they remain relevant as the architectural profession and society evolve.

(* Cf. Derived from Directive 85/384/EEC of the Commission of the European Communities)

Education

Definition:

Architectural education should ensure that all graduates have knowledge and ability in architectural design, including technical systems and requirements as well as consideration of health, safety, and ecological balance; that they understand the cultural, intellectual, historical, social, economic, and environmental context for architecture; and that they comprehend thoroughly the architects' roles and responsibilities in society, which depend on a cultivated, analytical and creative mind.

Background:

In most countries, architectural education is conventionally delivered by 4-6 years full-time academic education at a university (followed, in some countries, by a period of practical experience/training/internship), though historically there have been important variations (part-time routes, work experience etc.).

Policy:

In accordance with the UIA/UNESCO Charter for Architectural Education, the UIA advocate that education for architects (apart from practical experience/training/internship) be of no less than 5 years duration, delivered on a full-time basis in an accredited/validated/recognized architectural program in an accredited/validated/recognized university, while allowing variety in their pedagogic approach and in their responses to local contexts, and flexibility for equivalency.

Accreditation/Validation/Recognition

Definition:

This is the process that establishes that an educational program meets an established standard of achievement. Its purpose is to assure the maintenance and enhancement of an appropriate educational foundation.

Background:

Validated criteria and procedures for accreditation/validation/recognition by an independent organization help to develop well integrated and coordinated programs of architectural education. Experience shows that standards may be harmonized and promoted by regular, external monitoring, in some countries, in addition to internal quality assurance audits.

Policy:

That courses must be accredited/validated/recognized by an independent relevant authority, external to the university at reasonable time intervals (usually no more than 5-years), and that the UIA, in association with the relevant national organizations of higher education, develop standards for the content of an architect's professional education that are academically structured, intellectually coherent, performance-based and outcome-oriented, with procedures that are guided by good practice.

Practical Experience/Training/Internship

Definition:

Practical experience/training/internship is a directed and structured activity in the practice of architecture during architectural education and/or following receipt of a professional degree but prior to registration/licensing/certification.

Background:

To complement academic preparation in order to protect the public, applicants for registration/licensing/certification must integrate their formal education through practical training.

Policy:

That graduates of architecture will be required to have completed at least 2 years of acceptable experience/training/internship prior to registration/licensing/certification to practice as an architect (but with the objective of working towards 3 years) while allowing flexibility for equivalency.

Demonstration of Professional Knowledge and Ability

Definition:

Every applicant for registration/licensing/certification as an architect is required to demonstrate an acceptable level of professional knowledge and ability to the relevant national authority.

Background:

The public is assured of an architect's knowledge and ability only after he or she has acquired the requisite education and practical experience/training/internship, and demonstrated minimum knowledge and ability in the comprehensive practice of architecture. These qualifications have to be demonstrated by examination and/or other evidence.

Policy:

That the acquired knowledge and ability of an architect have to be proven by providing adequate evidence. This evidence must include the successful completion of at least one examination at the end of the practical experience/training/internship. Necessary components of professional practice knowledge and ability that are not subject to an examination have to be proven by other adequate evidence. These include such subjects as business administration and relevant legal requirements.

Registration/Licensing/Certification

Definition:

Registration/licensing/certification is the official legal recognition of an individual's qualification allowing her or him to practice as an architect, associated with regulations preventing unqualified persons from performing certain functions.

Background:

Given the public interest in a quality, sustainable built environment and the dangers and consequences associated with the development of that environment, it is important that architectural services are provided by properly qualified professionals for the adequate protection of the public.

Policy:

That the UIA promote the registration/licensing/certification of the function of architects in all countries. In the public interest, provision for such registration/licensing/certification should be by statute.

Procurement

Definition:

The process by which architectural services are commissioned.

Background:

Architects (through their codes of conduct) uphold the interests of their clients and society at large before their own interests. In order to ensure they have adequate resources to perform their functions to the standards required in the public interest, they are traditionally remunerated in accordance with either mandatory or recommended professional fee-scales.

There are international rules, such as the General Procurement Agreement (WTO) and the EU Services Directive, that aim to guarantee the objective and fair selection of architects. However, there has been an increasing tendency recently to select architects, for both public and private work, on the basis of price alone. Price-based selection forces architects to reduce the services provided to clients, which in turn compromises design quality and therefore the quality, amenity and social/economic value of the built environment.

Policy:

To ensure the ecologically sustainable development of the built environment and to protect the social, cultural, and economic value of society, governments should apply procurement procedures for the appointment of architects that are directed to the selection of the most suitable architect for projects. Conditional upon adequate resources being agreed among the parties, this is best achieved by one of the following methods:

- Architectural design competitions conducted in accordance with the principles defined by the UNESCO-UIA international competitions guidelines and approved by national authorities and/or architectural professional associations.
- A qualification based selection (QBS) procedure as set out in the UIA guidelines;
- Direct negotiation based on a complete brief defining the scope and quality of architectural services.

Ethics and Conduct

Definition:

A code of ethics and conduct establishes a professional standard of behavior that guides architects in the conduct of their practices. Architects should observe and follow the code of ethics and conduct for each jurisdiction in which they practice.

Background:

Rules of ethics and conduct have as their primary object the protection of the public, caring for the less powerful and the general social welfare, as well as the advancement of the interests of the profession of architecture.

Policy:

The existing UIA International Code of Ethics on Consulting Services remains in force. Member Sections of the UIA are encouraged to introduce into their own codes of ethics and conduct the recommended Accord Guidelines and a requirement that their members abide by the codes of ethics and conduct in force in the countries and jurisdictions in which they provide professional services, so long as they are not prohibited by international law or the laws of the architect's own country.

Continuing Professional Development

Definition:

Continuing Professional Development is a lifelong learning process that maintains, enhances, or increases the knowledge and continuing ability of architects.

Background:

More and more professional bodies and regulatory authorities require their members to devote time (typically at least 35 hours per year) to maintaining existing skills, broadening knowledge, and exploring new areas. This is increasingly important to keep abreast with new technologies, methods of practice, and changing social and ecological conditions. Continuing professional development may be required by professional organizations for renewal and continuation of membership.

Policy:

That UIA urge its member sections to establish regimes of continuing professional development as a duty of membership, in the public interest. Architects must be sure they are capable of providing the services they offer, and codes of conduct must oblige architects to maintain a known standard in a variety of areas described under the "Fundamental Requirements of an Architect" and in future variations thereof. In the meantime, the UIA must monitor the developments in continuing professional development for registration renewal, recommend guidelines among all nations to facilitate reciprocity and continue to develop policy on this subject.

Scope of Practice

Definition:

This is the provision of design and management services in connection with land-use planning, urban design, and building projects.

Background:

As society has evolved, the creation of the urban and built environment has become more complex. Architects have to deal with an increasingly wide range of urban, aesthetic, technical, and legal considerations. A coordinated approach to building design has proved to be necessary to ensure that legal, technical, and practical requirements are met and that society's needs and demands are satisfied.

Policy:

That the UIA encourage and promote the continuing extension of the boundaries of architectural practice, limited only by the provisions of codes of ethics and conduct, and strive to ensure the corresponding extension of the knowledge and skills necessary to deal with any extension of boundaries.

Form of Practice

Definition:

The legal entity through which the architect provides architectural services.

Background:

Traditionally, architects have practiced as individuals, or in partnerships or in employment within public or private institutions. More recently, the demands of practice have led to various forms of association, for example: limited and unlimited liability companies, cooperative practices, university-based project offices, community architecture, although not all are allowed in all countries. These forms of association may also include members of other disciplines.

Policy:

That architects should be allowed to practice in any form legally acceptable in the country in which the service is offered, but always subject to prevailing ethical and conduct requirements. The UIA, as it deems necessary, will develop and modify its policies and standards to take account of alternative forms of practice and varied local conditions where these alternatives are thought to extend the positive and creative role of the architectural profession in the interests of society.

Practice in a Host Nation

Definition:

Practice in a host nation occurs when an individual architect or corporate entity of architects either seeks a commission or has been commissioned to design a project or offer a service in a country other than his/her/its own.

Background:

There is an interest in increasing the responsible mobility of architects and their ability to provide services in foreign jurisdictions. There is also a need to promote the awareness of local environmental, social, and cultural factors and ethical and legal standards.

Policy:

Architects providing architectural services on a project in a country in which they are not registered shall collaborate with a local architect to ensure that proper and effective understanding is given to legal, environmental, social, cultural, and heritage factors. The conditions of the association should be determined by the parties alone in accordance with UIA ethical standards and local statutes and laws.

Intellectual Property and Copyright

Definition:

Intellectual property encompasses the three legal areas of patent, copyright, and trademark. It refers to the right (sometimes guaranteed under the law of some nation states) of designers, inventors, authors, and producers, to their ideas, designs, inventions, works of authorship, and the identification of sources of products and services.

Background:

While many countries have some legal protection covering the architect's design, that protection is often inadequate. It is not unusual for the architect to discuss ideas and concepts with a prospective client, subsequently not be hired, and later find that the client has used the architect's ideas with no recompense. The intellectual property of architects is, to some extent, protected by international regulations. In the context of the GATS, this is the agreement on trade-related aspects of intellectual property rights, including trade in counterfeit goods (TRIPS). The World Copyright Convention of September 16, 1955 is also of international significance. In Europe, the Revised Berne Agreement of 1886 is binding in most states.

Policy:

That the national law of a UIA member section should entitle an architect to practice his/her profession without detriment to his/her authority and responsibility, and to retain ownership of the intellectual property and copyright of his/her work.

Role of Professional Institutes of Architects

Definition:

Professions are generally controlled by a governing body that sets standards (e.g. of education, ethical rules, and professional standards to be observed). The rules and standards are designed for the benefit of the public and not the private advantage of the members. In some countries, certain types of work are reserved to the profession by statute, not in order to favor members but because such work should be carried out only by persons with requisite education, training, standards and discipline, for the protection of the public. Institutes have been established for the advancement of architecture, promotion of knowledge and – by ensuring that their members perform to a known standard – protection of the public interest.

Background:

Depending on whether a country has protection of title or function, (or both, or neither), the role and responsibilities of professional institutes varies considerably. In some countries, the statutory bodies also represent the profession; in others, these functions are separate.

It is customary for members of professional institutes to be expected to maintain a known standard. This is achieved by adhering to codes of conduct promulgated by the professional institutes, and fulfilling other requirements of membership, e.g. continuing professional development.

Policy

In countries where professional institutes do not exist, the UIA should encourage members of the architectural profession to form such institutes in the public interest.

Professional Institutes should seek to ensure that their members adhere to the UIA international standards, the minimum requirements of the UIA-UNESCO Architectural Education Charter, and UIA International Code of Ethics and Conduct; keep up to date their knowledge and skills as required by the list of "Fundamental Requirements" (both current, and as they evolve in the future); and generally contribute to the development of architectural culture and knowledge as well as the society they serve.

Appendix A

Resolution of Adoption (Number 17) of the UIA Accord on Recommended International Standards on Professionalism In Architectural Practice

Adopted by the XXI UIA Assembly
Beijing, China, July 28, 1999

The Assembly unanimously resolved that it adopts the Second Edition of the UIA Accord on Recommended International Standards of Professionalism in Architectural Practice as an advisory document intended to be used by member sections in setting and reviewing their own standards. The Accord and Guidelines will also make it easier for UIA member sections to negotiate mutual recognition agreements.

The Assembly asks that the Accord be transmitted to all UIA member sections with the request for their cooperation and participation in the further development of this policy framework for presentation at the XXII UIA Assembly (Berlin 2002).

The Assembly recognizes the mandate of Council to adopt Accord Policy Guideline documents and commend them to the UIA member sections.

The Assembly recognizes that there are differences in the cultures, practices and conditions in different member sections and encourages the member sections to use the documents as advisory documents intended to be adapted to local conditions.

The Assembly acknowledges that the sovereignty of each UIA member section must be respected in negotiations of mutual recognition agreements and notices that the guidelines are intended to allow flexibility for principles of equivalency and reciprocity and are structured to allow for the addition of requirements reflecting local conditions of a UIA member section.

The Assembly authorizes the UIA President and Secretary General to submit the Accord to the World Trade Organization, to other interested institutions and organizations as the basis for mutual recognition negotiations and to the Government of a country on the specific request of the UIA member section of the country in question.

The Assembly requests that the Professional Practice Commission analyze all the comments expressed during the General Assembly during its meeting in Prague (October 1999) in order to check on whether or not it is opportune to integrate them in the documents approved by the Beijing Assembly.

The Assembly authorizes the UIA Council to develop a policy to communicate the Accord and Guideline documents to interested parties.

The Assembly recommends to UIA member sections that following the use of these Standards, they inform the Commission Secretariat of their experience, in order that it can be taken into account for the improvement and evolution of these basic documents.

International Union of Architects

Recommended Guidelines for the UIA Accord On Recommended International Standards of Professionalism in Architectural Practice Policy on Accreditation/Validation/Recognition

March 29, 1998
Revised April 28, 1998
Revised December 10—12, 1998
Approved June, 1999

UIA Professional Practice Program Joint Secretariat

The American Institute of Architects
Co-Director James A. Scheeler, FAIA
1735 New York Avenue, NW
Washington, DC 20006
Telephone: 202 – 6267315
Facsimile: 202 – 6267421

The Architectural Society of China
Co-Director Zhang Qinnan, Vice President
Bai Wan Zhuang, West District
Beijing, China 100835
Telephone: 8610 – 88082239
Facsimile: 8610 – 88082222

Accord Policy on Accreditation/Validation/Recognition

That courses must be accredited/ validated/ recognized by an independent relevant authoritiy external to the university at reasonable time intervals (usually no more than five-years), and that the UIA, in association with the relevant national organizations of higher education, develop standards for the content of an architect's professional education that are academically structured, intellectually coherent, performance-based, and outcome-oriented, with procedures that are guided by good practice.

Introduction

Accreditation of educational programs in architecture, whether sought voluntarily by the educational institution or exacted by relevant authorities, seeks primarily to ensure, in the public interest, that the standards attained by succesful graduates of the program are adequate with regard to the design, technical, and professional skills and ethical formation required for competent architectural practice.

The principles in any accreditation policy permit flexibility of approach while ensuring independent standards for the accrediting body and the pursuit and maintenance of high standards in educational endeavor and in the accrediting process itself.

The critical criteria in a satisfactory educational program involve thorough assessment in accordance with previously defined and agreed criteria, by a group of assessors external to the school of architecture who are competent by training and experience to evaluate architecture programs and make recommendations for their direction or modification. External assessors may be appointed by the state in which the architectur program is run, by an independent professional architectural organization, nominated as external examiners by the school of architecture itself, or by some other satisfactory method. The system of appointing the assessor may vary depending upon whether the educational institution is publicly or privately run. An independent relevant authoritiy can comprise representatives of a professional body, such as an institute or chamber of architects, or a nongovernmental organization of architects or schools of architecture; it can be a national or state government, or its delegated representatives, or an organization of external examiners. The process of validation of programs will occur periodically, and a satisfactory method of accreditation will involve the review of the work of all students passing through a school of architecture on at least one occasion during their educational program. Recognizing the differences between governance of public and private institutions, consistency must be provided both in the accreditation/recognition/validation process and the end result.

Accreditation procedures vary depending on whether the education programs in question are proposed for establishment, recently set up and not previously accredited, or the subject of proposed change. In every instance, the assessors will be provided with documentation in advance of their visit to the school; review examination papers and scripts, studio programs and studio work; course syllabus and examples of course work; and meet with students and staff. They may also look at the pedagogic, professional, and research output of the faculty/staff. On conclusion of such review, the assessors will provide the school with a program report that will make recommendations for accreditation and may make suggestion for changes in the educational program or impose conditions for accreditation.

1. Criteria for Accreditation Courses, Programs, and Examinations in Architecture

The core knowledge and skills required of a competent architect, set down by the relevant organizations for higher education and recognized in the UIA Accord on Recommended International Standards of Professionalism in Architectural Practice, are as set down in the Fundamental Requirements of an Archtitect from the Accord.

These skills are mastered by the architect through education, training, and experience, and educational programs in architecture set out to help the student of architecture acquire such ability, knowledge, unterstanding, and skills to the extent that these may be required within such a program.

The UIA advocates that education for architects should be of no less than five years duration, principally delivered on a full-time basis in an accredited architectural program in an accredited/validated university, while allowing flexibility for equivalency. In some countries, education is followed by a period of practical experience/training/internship. During this education and training process, the levels of ability reached by the student of architecture in the fundamental requirements listed will advance in line with the progress of the student's study, and validation of the relevant educational program will take account of the varying levels of attainment reasonably to be expected at the appropriate moments.

The knowledge and abilities required of architects have changed and will continue to change to reflect society's expectations. The UIA will review its Recommended Guidelines for the Accord Policy on Accreditation/ Validation/ Recognition from time to time to ensure its continual relevance.

The relative weighting ascribed to the different criteria listed and the relative degree of importance of skill to be attained will vary from country to country and from time to time. In different countries, for reasons of tradition and deliberate choice, educational institutions may themselves ascribe different weighting to the various criteria, which, in turn, will be influenced by the precise missions generally undertaken by architects within that country. In every instance, the educational program will be based on a syllabus that will incorporate topics and subjects derived from or comparable to those listed. The accreditation criteria will include a review of the syllabus. The syllabus will vary depending on the stage of studies, whether at intermediate or final examination level and whether before or after any period of practical training.

2. Methods of Accreditation

Accreditation is carried out by properly constituted authorities that are independent of the institution housing the program to be accredited. Accrediting authorities must be competent by way of training and experience. This will indicate that persons undertaking accreditation work have experience in architectural design, practice, ethical standards, and training. Frequently an accreditation panel will comprise nominees of more than one of the types of organizations listed and, in every instance, involves accreditation by established members of the architectural profession. This will help promote both objective evaluation and a broad and inclusive view of architecture.

In every instance, when educational institutions participate in accreditation procedures, the educational institiution cannot participate in the procedure for accrediting its own program.

3. Procedures for Accrediting Educational Programs in Architecture

The nature and detail of procedures to be adopted by an accrediting board will vary depending on the culture and educational practices of the country concerned. They will also vary on whether an educational program is being considered in advance of its establishment; examined for the first time; has been established for some time and has previously been accredited; or, having either failed to achieve accreditation or having had a previous accreditation withdrawn, is presented for accreditation afresh for a further time.

Accreditation procedures will also vary depending on whether one or more stages in the process are to be accredited. In some countries, accreditation procedures involve a three-stage process: after three and five years respectively in the academic educational program and on conclusion of an agreed period of practical training. In other countries the process will involve one or two stages.

Accreditation procedures involve the review by the assessors of the content of an educational program and of the standards achieved within that program. The assessment is made on the basis of information provided by the educational institution with regard to the program, syllabus, details of studio programs and examination scripts, and reports of external examiners; a self-appraisal by the educational institution; and, during a visit to the institution, on meetings with the head of the school program staff and students and inspection of student work and facilities.

Where an institution is proposing major changes to an existing course or proposing to introduce a new course, it may be helpful to undertake a preliminary assessment by an independent relevant authority as to whether the content, structure, and resources of the proposed program are such as to be likely to achieve accreditation of the course and its graduates. Information that will be useful in such an assessment would include a description of the context of the proposal, philosophical approach proposed, and proposed academic program. Such a description might include a course diagram, details of the course framework, requirements to complete the course, and details of lecture syllabi and contact hours for each subject.

4. Documentation and Visiting Methods

Where accreditation is being sought either on an initial or ongoing basis for an already-established educational program, documentation to be provided by the educational institution to the accrediting authority might include:

- A brief description of the parent educational institution, with a statement of factors within the national, regional, and urban context that influence the nature of the educational institution;
- A brief description of the history of the course;
- The philosophical approach, mission, and vision to architectural education;
- An indication of any characteristics in the background of students that influence the direction of courses offered;
- A summary of academic staff profile, including nonteaching activities and other duties including research, publications, professional work, and community involvement;
- A statement of physical resources, including studios, teaching space and equipment, laboratory and workshops, library facilities and resource centers, computers, and information systems;
- A note of decision-making networks and management structure;
- A complete description of the academic program, including a description of the program framework, requirements to complete the program, and other requirements for graduation; lecture syllabi; details of studio programs; and copies of course handbooks;
- Statistical information on student enrollment numbers, numbers of graduations, staff numbers, and the staff/student ratio;
- A self-appraisal by the school of its education policy — taking account, where appropriate, of reports provided by previous accreditation boards and discussing development since any previous accreditation — to cover issues in external examiners' reports, changes in resource provisions, critical evaluaton of course objectives, special features of the course, and other relevant matters.

The accrediting authority visits the educational institution and reviews the program *in situ*. During the visit, an exhibition of work completed by students over a period of at least 12 months prior to the visit will be helpful. Such exhibition should comprise a range of studio work, with programs attached for each year of the course arranged as far as possible to show the development of the curriculum throughout the program. Arange of the written and drawn work in each year of the program should be exhibited so that the level of attainment of students in each of the areas as set out as fundamental requirements for an architect can be assessed. Presented work should include a representative sample of studio portfolios and examination scripts for the highest, average. and lowest pass grades in each subject, and these should be complemented by records of examination and assessment results for all years of the course in all subject areas.

When inspecting the educational program *in situ*, the accrediting authority may wish to untertake meetings and discussions with the program teachers, including the head of the school or department, studio and specalist staff, and external examiners. The authority may also talk with students of the program, both as a body and/or individually. Subject for discussion as part of the assessment process might include methods of educational assessment; the content of project work and lecture courses; and the relation of lecture courses to project work, the use of specialist teachers, and future developments.

5. Reporting Procedures

An accrediting authoritiy will provide a written report on the educational program on conclusion of the visit to the program. Such a report will validate and supplement the written information provided by the educational institution and convey the accrediting authority's view of the quality of education in terms of student performance in the course under review. Procedures might include methods of ensuring such report is free from factual error, is treated confidentially, and is seen by all relevant parties. An accrediting report will normally recommend accreditation of the educational program for a fixed period of no more than five years' duration.

International Union of Architects

Recommended Guidelines for the UIA Accord On Recommended International Standards of Professionalism in Architectural Practice Policy on Practical Experience/Training/Internship

April 1998
Revised December 10—12, 1998
Adopted June, 1999

∽∽∽∽∽∽∽∽∽∽∽∽∽∽∽∽∽∽∽∽∽∽∽∽∽∽∽∽∽∽

UIA Professional Practice Program Joint Secretariat

The American Institute of Architects	The Architectural Society of China
Co-Director James A Scheeler, FAIA	Co-Director Zhang Qinnan, Vice President
1735 New York Avenue, NW	Bai Wan Zhuang, West District
Washington, DC 20006	Beijing, China 100835
Telephone: 202 - 6267315	Telephone: 8610 - 88082239
Facsimile: 202 - 6267421	Facsimile: 8610 - 88082222

Accord Policy on Practical Experience/ Training/ Internship

That graduates of architecture will be required to have completed at least two years of acceptable experience/ training/ internship prior to registration/ licensing/ certification to practice as an architect (but with the objective of working towards three years) while allowing flexibility for equivalency.

Guidelines

1. Period of Practical Experience/Training/Internship

The experience set out below should be demonstrated prior to applying for registration/licensing/certification and should be gained over the period defined in the Accord Policy. At least half of that period should occur following the basic academic prerequisites and in any case should not imply a reduction of the academic period referred to under the Accord Education Policy.

2. Objectives of The Period of Practical Experience/Training/Internship

The objectives of the period of practical experience/training/internship (here after referred to as internship) are:
- To provide interns with the opportunity to acquire basic knowledge and skill in the practice of architecture;
- To ensure the practices, activities, and experience of interns is recorded by a standard method;
- To enable interns to attain a broad range of experience in the practice of architecture.

3. Categories of Experience

An intern should receive practical experience and training under the direction of an architect in at least half of the areas of experience nominated under each of the following four categories:

3.1 Project and Office Management
Meeting with clients
Discussions with clients of the brief and the preliminary drawings

Formulation of client requirements
Pre-contract project management
Determination of contract conditions
Drafting of correspondence
Coordination of the work of consultants Office and project accounting systems
Personnel issues

3.2 Design and Design Documentation
Site investigation and evaluation
Meetings with relevant authorities
Assessment of the implications of relevant regulations
Preparation of schematic and design development drawings
Checking design proposals against statutory requirements
Preparation of budgets, estimates, cost plans, and feasibility studies

3.3 Construction Documents
Preparation of working drawings and specifications
Monitoring the documentation process against time and cost plans
Checking of documents for compliance with statutory requirements
Coordination of subcontractors documentation
Coordination of contract drawings and specifications

3.4 Contract Administration
Site meetings
Inspection of works
Issuing instructions, notices, and certificates to the contractor
Client reports
Administration of variations and monetary allowances

4. Record of Practical Experience/Training/Internship
Interns should maintain a written record, in a standard form or a logbook, of all periods of training, experience, and supplementary education received during the internship period.
This record should be set out under the areas of training nominated in Part 2 above. It should describe the nature and duration of activities undertaken, and each of these should be signed by the supervisor architect as a true record of the experience gained by the intern.

The standard form or logbook is to be presented to the registering/licensing authority on request, as evidence that the required practical experience/ training/ internship is being undertaken or has been completed.

5. Supervisors

Interns should gain their experience under supervision. Supervisors are to be registered or licensed architects in the jurisdiction in which the internship is undertaken, and will either be the employer or the architect to whom the intern reports during each recorded period of experience.

6. Core Knowledge and Ability Requirements

At the completion of the period of practical experience/training/internship, the intern should have demonstrated or be able to demonstrate knowledge and/or ability in the following:

6.1 The Practice of Architecture
- An overview of the architectural profession in the national and international community
- A knowledge and appreciation of ethical standards
- Knowledge of the local architectural association
- An overview of the local construction industry and construction law
- Direction and coordination of consultants
- Office management and systems
- Legal aspects of practice
- Liability, risk management, and insurance

6.2 Project Management
- Establishing and managing client agreements
- Scheduling of project activities and tasks
- Assessing codes, regulations, and legislation
- Project financing and cost control
- Project procurement and contractual systems
- Dispute resolution
- Management of subcontractors
- Project administration and monitoring systems

6.3 Pre-design and Site Analysis

- Establishing, analyzing, and recording environmental issues relevant to the project
- Establishing and clearly defining a design brief
- Establishing, analyzing, and recording site conditions

6.4 Project Services and Systems
- Coordinating the design and documentation of project services and systems into the project design and documentation process

6.5 Schematic Design
- Analyzing the client brief and producing potential project design solutions through a process of hypothesis, evaluation, and reappraisal
- Graphically representing alternative project designs
- Presenting and agreeing preliminary design proposals with clients and other interested parties

6.6 Design Development and Design Documentation
- Investigating and establishing the specific spatial, organization and circulation requirements within and around a project
- Considering and deciding upon the disposition of construction and project services systems, materials, and components
- Developing drawings and documents to fully describe the developed design proposal for the approval of the client and other interested parties
- Analyzing possible effects on the context, users, etc.

6.7 Construction Documentation
- Researching, analyzing, and selecting appropriate materials and systems for a roject
- Preparing accurate consistent and complete construction drawings, specifications, and schedules that describe the extent and location of construction elements, components, finishes, fittings, and systems

6.8 Contract Administration
- Preparing documents to invite bids or tenders
- Evaluating and making recommendations in respect of bids or tenders received
- Finalizing project contracts
- Administering project contracts

- Monitoring compliance with contract conditions and the requirements of relevant authorities
- Inspecting and evaluating construction works to ensure that they comply with the requirements of the contract documents

International Union of Architects

Recommended Guidelines for the UIA Accord On Recommended International Standards of Professionalism in Architectural Practice Policy on Registration/Licensing/Certification of the Practice of Architecture

September 5, 1997
Revised March 4, 1998
Revised April 17, 1998
Revised December 10—12, 1998
Adopted June, 1999

UIA Professional Practice Program Joint Secretariat

The American Institute of Architects
Co-Director James A. Scheeler, FAIA
1735 New York Avenue, NW
Washington, DC 20006
Telephone: 202 - 6267315
Facsimile: 202 - 6267421

The Architectural Society of China
Co-Director Zhang Qinnan, Vice President
Bai Wan Zhuang, West District
Beijing, China 100835
Telephone: 8610 - 88082239
Facsimile: 8610 - 88082222

Accord Policy on Registration/Licensing/Certification of the Practice of Architecture

That the UIA promote the registration/ licensing/ certification of the function of architects in all countries. In the public interest, provision for such registration/ licensing/ certification should be by statute.

Introduction

Registration/Licensing
Certification

Registration/licensing/certification is the official legal recognition of an individual's qualification allowing her or him to practice as an independent architect, associated with regulations preventing unqualified persons from performing certain functions. Given the public interest in a high-quality, sustainable built environment and the dangers and consequences associated with the construction industry, it is important that architectural services are provided by properly qualified professionals in order to provide adequate protection for the public.

Registration/licensing/certification is based on minimum standards of competency relative to education, experience, and examination to ensure that the public interest is served. Occupational licensure is an exercise of the state's inherent police power to protect the health, safety, and welfare of its citizens. Five generally accepted criteria indicate when licensure is appropriate: 1) unregulated practice of the occupation poses a serious risk to a consumer's life, health, safety, or economic well-being and the potential for harm is recognizable and likely to occur; 2) the practice of the occupation requires a high degree of skill, knowledge, and training; 3) the functions and responsibilities of the practitioner require independent judgment and the members of the occupational group practice independently; 4) the scope-of-practice of the occupation is distinguishable from other licensed and unlicensed occupations; 5) the economic and cultural impact on the public of regulating this occupational group is justified. The practice of architecture meets these classic criteria.

Practice Regulation vs. Title Registration

"**Practice regulation**" i.e. regulation of the practice of a profession, means that only those individuals who meet specific legislated criteria (of education, training, and testing) may perform the services of a profession.

Practice regulation or licensure—because of its cost to the state and consumers and because it limits entry into a profession—is traditionally reserved for professions and occupations that if unregulated pose a serious threat to public health, safety, and welfare. In evaluating whether a profession should be regulated by practice regulation, most states apply a set of objective criteria, which include: Is the public being harmed by lack of regulation and can such harm be documented? Are there alternatives to state regulation? Is the public protected by existing laws, codes, or standards, and would strengthening such laws solve the problem? What is the cost to the state and the public of regulating the profession and will the public benefit from such regulation?

"**Title registration**" means individuals must still meet specific qualifications criteria, but only the use of the title is controlled. Individuals who do not have the title may continue to perform the services. Title registration should confer only a protected title. A title bill should not affect the scope of that group's practice or permit those individuals to do anything they were not already legally entitled to do. (NOTE: Title registration is called "certification" in most states. The word "licensing," though often used as an umbrella term for state regulation, is used by most states to mean practice regulation.)

Title registration is intended to provide a means that the public can use for distinguishing trained/qualified practitioners or providers of a service from untrained or unqualified individuals. Title registration does not prevent other less qualified individuals from providing the services; it simply establishes a measuring stick against which their qualifications can be judged. Title registration is considered appropriate when no serious threat to the public is involved, but consumers may be confused and misled about providers' qualifications.

Title registration achieves the goal of enabling the public and consumers of services to differentiate, with minimal cost to the state and consumers, trained, qualified individuals from those who are untrained. With title registration, those individuals who do not meet the registration requirements are not deprived of their livelihoods. These individuals can continue to provide services; they are simply constrained from using a protected title.

Proposed Legislative Guidelines

The International Union of Architects recommends that legislation or statutes regulating the profession of architecture should be based on regulation of the practice of architecture. The following guidelines reflect that recommendation and set forth provisions to deal with a limited number of problem areas of state regulation that have implications beyond the boundaries of an individual state. For the sake of brevity, the term "registration" is used throughout the guideline to denote "registration/ licensing/ certification." It should be noted that in any mutual recognition agreement between national and international jurisdictions, the UIA takes the position that only registered architects (whether with practice registration or title registration) are recognized.

Guidelines rather than draft statutory language are recommended because the laws of states represented by the member sections of the UIA contain language, organization, and provisions reflecting the unique political and cultural characteristics of those states. It would undoubtedly be disruptive and confusing to try to suggest exact statutory language on an international basis.

1. Definition

1.1 Practice of Architecture: For the purpose of a registration statute, the definition of the practice of architecture should be the definition adopted by the UIA in the Accord on Recommended International Standards of Professionalism in Architectural Practice:

> *The practice of architecture consists of the provision of professional services in connection with town planning and the design, construction, enlargement, conservation, restoration, or alteration of a building or group of buildings. These professional services include, but are not limited to planning; strategic and land-use planning; urban design; provision of preliminary studies, designs, models, drawings, specifications, and technical documentation; coordination of technical documentation prepared by others as appropriate and without limitation (consulting engineers, landscape architects, and other specialist consultants); construction economics; contract administration;*

monitoring of construction (referred to as supervision in some countries); and project management.

This definition of the practice of architecture covers the wide variety of services that architects normally furnish and for which they are specifically trained and in which they are required to demonstrate professional competency. In some jurisdictions where the education and training and competency standards are more narrowly drawn, the UIA Accord definition may need to be amended to reflect these narrower standards.

No person should be permitted to engage in the practice of architecture unless registered or otherwise permitted to practice under the registration statute. No person should be permitted to use the title "architect" or otherwise represent to the public that he or she is an architect unless he or she is registered to practice architecture.

In some instances, state statutes may exempt various categories of related design professionals from the purview of the statute to the extent that the exercise of their profession may incidentally involve them in the practice of architecture. It is important that these exemptions be carefully thought out to serve as a means for setting off other legitimate design activities from the practice of architecture.

In many jurisdictions, engineering registration laws permit the engineer to design structures as well as a multitude of other projects. The architectural profession is often restricted by law to designing only buildings and ancillary facilities "for human habitation." The UIA advocates that statutes regulating the profession of architecture should not unduly narrow the scope of practice and should recognize that architects, through their practices, express the roots of a society's cultural and aesthetic values through the architecture they design.

2. Regulation of Conduct of Registrants

2.1 Authorization: Clearly the authorization of an architectural registration agency to adopt rules or regulations governing the conduct of architects should be covered by statute. Rule-making power coupled with a power of revocation or suspension of registration based on misconduct implicitly requires further description by the rule-making process of what will constitute misconduct.

2.2 Rules of Conduct: The statute should authorize the architectural registration agency to promulgate, as part of its regulatory function, rules of conduct governing the practice of registered architects. The statute should contain standards for the scope and content of such rules. The statute should also provide that violation of the rules of conduct promulgated by the architectural registration agency is one of the enumerated grounds for revocation or suspension of registration or for the imposition of a civil fine.

3. Qualification for Registration

Qualification criteria for registration should be objective and transparent. For the purpose of a registration statute, care should be taken to assure that the statute appropriately reflects the "UIA Accord on Recommended International Standards of Professionalism in Architectural Practice" policies and guidelines on fundamental requirements of an architect, education, accreditation/ validation/ recognition, practical experience/ training/internship, and practical examination/demonstration of professional knowledge and ability. It is not appropriate that statutes contain requirements for citizenship or residency to enter into the profession.

3.1 Degree: An applicant for registration should be required to hold an accredited professional degree in architecture. The UIA recommends that the UIA/UNESCO Charter for Architectural Education be established as the minimum criteria for architectural education.

3.2 Practical Training: The UIA recommends that an applicant for registration have such practical training as set out in the Accord Policy.

3.3 Examination: To be registered, the applicant should be required to pass examinations covering such subjects and graded on such basis as the registration agency shall, by regulations, decide.

3.4 Personal Interview: Registration agencies may require a personal interview with a candidate for registration.

3.5 Moral Character: If the state wishes to invest its registration agency with discretion to reject an applicant who is not of "good moral character," the statute should specify only the aspects of the applicant's background germane to the inquiry, such as:

- Conviction for commission of a felony;
- Misstatement or misrepresentation of fact by the applicant in connection with his or her application;
- Violation of any of the rules of conduct required of registrants and set forth in the statutes or regulations;
- Practicing architecture without being registered in violation of registration laws of the jurisdiction in which the practice took place.

If the applicant's background includes any of the foregoing, the registration agency should be allowed, notwithstanding, to register the applicant on the basis of suitable evidence of reform.

4. Reciprocity Procedure

The statute should make provision for registering nonresident applicants in addition to the provisions outlined in the section 3, Qualifications for Registration, and to any provisions in the statute providing other forms of reciprocity.

4.1 Nonresident Applicant Seeking to Practice: Every nonresident applicant seeking to practice architecture in a jurisdiction should be registered, if the applicant:

- Holds a current and valid registration issued by a registration authority recognized by mutual recognition agreement by the jurisdiction;
- Files an application with the jurisdiction, on a form prescribed by the jurisdiction, containing such information satisfactory to the jurisdiction concerning the applicant as the jurisdiction considers pertinent.

4.2 Nonresident Applicant Seeking a Commission: A nonresident applicant seeking an architectural commission in a jurisdiction in which he or she is not registered should be admitted to the jurisdiction for the purpose of offering to render architectural services and for that purpose only without having first been registered by the jurisdiction, if the applicant:

- Holds a current and valid registration issued by a registration authority recognized by mutual recognition agreement by the jurisdiction;
- Notifies the board of the jurisdiction in writing that (a) he or she holds a cur-

rent valid registration issued by a registration authority recognized by mutual recognition agreement by the jurisdiction but is not currently registered in the jurisdiction and will be present in the jurisdiction for the purpose of offering to render architectural services, (b) he or she will deliver a copy of the notice referred to in (a) to every potential client to whom the applicant offers to render architectural services, and (c) he or she shall apply immediately to the board for registration if selected as the architect for a project in the jurisdiction.

The applicant should be prohibited from actually rendering architectural services until he or she has been registered.

4.3 Design Competition: A person seeking an architectural commission by participating in an architectural design competition for a project in a jurisdiction in which he or she is not registered should be permitted to participate in the competition, if the person:

- Holds a current and valid registration issued by a registration authority recognized by mutual recognition agreement by the jurisdiction;
- Notifies the jurisdiction in writing that he or she is participating in the competition and holds a current and valid registration issued by a registration authority recognized by mutual recognition agreement by the jurisdiction;
- Undertakes to apply to the jurisdiction for registration immediately on being chosen as an architect for the project.

5. Form of Practice

If architectural services are provided by corporate entities, they should be required to be under the effective control of architects and required to conform to and maintain the same professional standards of service, work, and conduct as individual architects.

A majority of member sections responding to the UIA Professional Practice Commission questionnaire indicated that their states permitted the practice of architecture in partnerships and conventional corporate forms. The restrictions placed on corporate practice and the newer limitedliability company are often onerous. The great variety of these restrictions suggest that a guideline is needed to seek a reasonable, international provision respecting firm practice while assuring the public of the integrity of architectural services performed.

5.1 Practice Structure: The UIA guidelines recommend that statutes provide that a partnership (including a registered limited liability partnership), a limited liability company, or a corporation should be admitted to practice architecture in a jurisdiction if:

- At least two-thirds of the general partners, if a partnership; or two – thirds of the directors, if a limited liability company or a corporation, are registered under the laws of any state or country to practice architecture;
- The person having the practice of architecture in her or his charge is herself or himself a general partner, if a partnership; a director, if a limited liability company; or a director, if a corporation, and registered to practice in that jurisdiction.

The statute should empower the registration agency to require, by regulations, any partnership, limited liability or unlimited company, or corporation practicing architecture in that state to file information concerning its officers, directors, managers, beneficial owners, and other aspects of its business organization on such forms as the agency prescribes.

5.2 Firm Name: A firm otherwise qualified to practice in a state or country should be permitted to practice in that state or country under a name that does not include the names of every director, if a corporation; every manager, if a limited liability company; or every general partner, if a partnership, registered in any state or country to practice architecture, provided the firm complies with reasonable regulations of the registration agency requiring the firm to file the names, addresses, and other pertinent information concerning the directors, managers, or general partners of the firm.

6. Engagement of an Architect During Construction of a Project

Construction administration services, including periodic site visits, shop drawing review, and reporting violations of codes or substantial deviations from the contract documents constitute an important responsibility of the architect and assure the public health, safety, and welfare. The following guidelines are intended to ensure that at least the minimum of construction services are provided by the design architect:

6.1   An owner who proceeds to have constructed a project having as its principal purpose human occupancy or habitation shall be deemed to be engaged herself or himself in the practice of architecture unless she or he has employed an architect to perform at least minimum construction administration services, including periodic site visits, shop drawing review, and reporting to the owner and building official any violations of codes or substantial deviations from the contract documents that the architect observed.

6.2   It shall be the project design architect's obligation to report to the registration jurisdiction and to the building official if he or she is not engaged to provide construction administration services described in Paragraph 1, above.

6.3   A registration jurisdiction may waive these requirements with respect to a particular project or class of projects if it determines that the public is adequately protected without the necessity of an architect performing the services described in Paragraph 6.1.

7. Regulation of Unregistered Persons Practicing Architecture

The unregistered practice of architecture can endanger the public health, safety and welfare. The following guidelines provide a basis and means for enforcing the statute:

7.1   Although violation of the architectural registration statute by unregistered persons should be a crime, the registration agency should also be authorized, after a hearing, to impose civil fines of up to a stated amount and to issue orders to cease against unregistered persons and persons aiding and abetting unregistered persons. The registration agency as well as the government's attorney general and other local law enforcement authorities, should be authorized to seek injunctions against practice by unregistered persons and the aiding or abetting of such practice, and judicial enforcement of civil fines imposed by the registration agency.

7.2   All plans, specifications, and other technical submissions prepared in the course of practicing architecture (as defined in Guideline 1) required to be filed with the state of local building or public safety officials should be sealed by an architect. If state law provides certain exceptions to the general requirement that technical submissions be sealed, then the person filing the technical submissions should specify on them the state law exempting the preparation of those technical submissions. Any permit issued on the basis of technical submissions not complying with these requirements shall be invalid.

International Union of Architects

Recommended Guidelines for the UIA Accord On Recommended International Standards of Professionalism in Architectural Practice Policy on Ethics and Conduct

November 1997
Revised April 1998
Revised December 10—12, 1998
Adopted June, 1999

∽∽∽∽∽∽∽∽∽∽∽∽∽∽∽∽∽∽∽∽∽∽∽∽∽∽∽∽∽∽

UIA Professional Practice Program Joint Secretariat

The American Institute of Architects	The Architectural Society of China
Co-Director James A. Scheeler, FAIA	Co-Director Zhang Qinnan, Vice President
1735 New York Avenue, NW	Bai Wan Zhuang, West District
Washington, DC 20006	Beijing, China 100835
Telephone: 202 – 6267315	Telephone: 8610 – 88082239
Facsimile: 202 – 6267421	Facsimile: 8610 – 88082222

Accord Policy on Ethics and Conduct

The existing UIA International Code of Ethics on Consulting Services remains in force. Member Sections of the UIA are encouraged to introduce into their own codes of ethics and conduct the recommended Accord Guidelines and a requirement that their members abide by the codes of ethics and conduct in force in the countries and jurisdictions in which they provide professional services, so long as they are not prohibited by international law or the laws of the architect's own country.

Recommended Guidelines for the Accord Policy on Ethics and Conduct

Introduction

At the meeting of the commission in Washington in December, 1998, there was broad agreement that the amended code evolving from the Barcelona meeting should be put to the Assembly in Beijing for adoption as the Accord Guidelines for Ethics and Conduct for subsequent adoption by member sections within their own codes.

The drafting panel, drawing on principles and policies articulated in the accord and the codes of ethics and conduct from member sections around the world, recommend to the council and assembly the following:

Preamble

Members of the architectural profession are dedicated to the highest standards of professionalism, integrity, and competence, and to the highest possible quality of their output, and thereby bring to society special and unique knowledge, skills, and aptitudes essential to the development of the built environment of their societies and cultures. The following are principles for the conduct of architects in fulfilling those obligations when undertaking a consulting service. They apply to all professional activities, wherever they occur. They address responsibilities to the public, which the profession serves and enriches; to the clients and users of architecture and the building industries, who help to shape the built environment; and to the art and science of architecture, that continuum of knowledge and creation which is the heritage and legacy of the profession and of society.

Principle 1
General Obligations

Architects possess a systematic body of knowledge and theory of the arts, science, and business of architecture developed through education, training, and experience. The process of architectural education, training, and examination is structured to assure the public that, when an architect is appointed to perform professional services, that architect has met acceptable standards enabling proper performance of those services. Architects have a general obligation to maintain and advance their knowledge of the art and science of architecture, respect the body of architectural Accomplishment and contribute to its growth, and give precedence to learned and uncompromised professional judgement over any other motive in the pursuit of the art, science, and business of architecture.

1.1 Standard: Architects shall strive to continually improve their professional knowledge and skill in areas relevant to their practices.

1.2 Standard: Architects shall continually seek to raise the standards of aesthetic excellence, architectural education, research, training, and practice.

1.3 Standard: Architects shall, as appropriate, promote the allied arts and contribute to the knowledge and capability of the building industries.

1.4 Standard: Architects shall ensure that their practices have appropriate and effective internal procedures, including monitoring and review procedures, and sufficient qualified and supervised staff such as to enable them to function efficiently.

1.5 Standard: Where work is carried out on behalf of an architect by an employee or by anyone else acting under an architect's direct control, the architect is responsible for ensuring that that person is competent to perform the task and, if necessary, is adequately supervised.

Principle 2
Obligations to the Public

Architects have obligations to the public to embrace the spirit and letter of the laws governing their professional affairs, and should thoughtfully consider the social and environmental impact of their professional activities.

2.1 Standard: Architects shall respect and help conserve the systems of values and the natural and cultural heritage of the community in which they are creating architecture. They shall strive to improve the environment and the quality of the life and habitat within it in a sustainable manner, being fully mindful of the effect of their work on the widest interests of all those who may reasonably be expected to use or enjoy the product of their work.

2.2 Standard: Architects shall neither communicate nor promote themselves or their professional services in false, misleading or deceptive manners.

2.3 Standard: An architectural firm shall not represent itself in a misleading fashion.

2.4 Standard: Architects shall uphold the law in the conduct of their professional activities.

2.5 Standard: Architects shall abide by the codes of ethics and conduct and laws in force in the countries and jurisdictions in which they provide or intend to provide professional services.

2.6 Standard: Architects shall as appropriate involve themselves in civic activities, as citizens and professionals, and promote public awareness of architectural issues.

Principle 3
Obligations to the Client
Architects have obligations to their clients to carry out their professional work faithfully, conscientiously, competently, and in a professional manner, and should exercise unprejudiced and unbiased judgement with due regard to the relevant technical and professional standards when performing all professional services. Learned and professional judgement should take precedence over any other motive in the pursuit of the art, science, and business of architecture.

3.1 Standard: Architects shall only undertake professional work where they can ensure that they possess adequate knowledge and abilities and where adequate financial and technical resources will be provided in order to fulfil their commitments in every respect to their clients, for any one commission.

3.2 Standard: Architects shall perform their professional work with due skill care and diligence.

3.3 Standard: Architects shall carry out their professional work without undue delay and, so far as it is within their powers, within an agreed reasonable time limit.

3.4 Standard: Architects shall keep their client informed of the progress of work undertaken on the client's behalf and of any issues that may affect its quality or cost.

3.5 Standard: Architects shall accept responsibility for the independent advice provided by them to their clients, and undertake to perform professional services only when they, together with those whom they may engage as consultants, are qualified by education, training, or experience in the specific areas involved.

3.6 Standard: Architects shall not undertake professional work unless the parties have clearly agreed in writing to the terms of the appointment, notably:

- Scope of work;
- Allocation of responsibilities;
- Any limitation of responsibilities;
- Fee or method of calculating it;
- Any provision for termination.

3.7 Standard: Architects shall be remunerated solely by the fees and benefits specified in the written agreement of engagement or employment.

3.8 Standard: Architects shall not offer any inducements to procure an appointment.

3.9 Standard: Architects shall observe the confidentiality of their client's affairs and should not disclose confidential information without the prior consent of the client or other lawful authority; for example, when disclosure is required by order of a court of law.

3.10 Standard: Architects shall disclose to clients, owners, or contractors significant circumstances known to them that could be construed as creating a conflict of interest, and should ensure that such conflict does not compromise the legitimate interests of such persons or interfere with the architect's duty to render impartial judgement of contract performance by others.

Principle 4
Obligations to the Profession

Architects have an obligation to uphold the integrity and dignity of the profession, and shall in every circumstance conduct themselves in a manner that respects the legitimate rights and interests of others.

4.1 Standard: Architects shall pursue their professional activities with honesty and fairness.

4.2 Standard: An architect shall not take as a partner and shall not act as a co-director with an unsuitable person, such as a person whose name has been removed from any register of architects otherwise than at his own request or a person disqualified from membership of a recognised body of architects.

4.3 Standard: Architects shall strive, through their actions, to promote the dignity and integrity of the profession, and to ensure that their representatives and employees conform their conduct to this standard, so that no action or conduct is likely to undermines the confidence of those for and with whom they work and so that members of the public dealing with architects are protected against misrepresentation, fraud, and deceit.

4.4 Standard: Architects shall, to the best of their ability, strive to contribute to the development of architectural knowledge, culture, and education.

Principle 5
Obligations to Colleagues

Architects should respect their rights and acknowledge the professional aspirations and contributions of their colleagues and the contribution made to their works by others.

5.1 Standard: Architects shall not discriminate on grounds of race, religion, disability, marital status, or gender.

5.2 Standard: Architects shall not appropriate the intellectual property of nor unduly take advantage of the ideas of another architect without express authority from the originating architect.

5.3 Standard: Architects shall not, when offering services as independent consultants, quote a fee without receiving an invitation to do so. The must have sufficient information on the nature and the scope of the project to enable a fee proposal to be prepared that clearly indicates the service covered by the fee in order to protect the client and society from unscrupulous under-resourcing by an architect.

5.4 Standard: Architects shall not, when offering services as independent consultants, revise a fee quotation to take account of the fee quoted by another architect for the same service in order to protect the client and society from unscrupulous under-resourcing by an architect.

5.5 Standard: The architect shall not attempt to supplant another architect from an appointment.

5.6 Standard: Architects shall not enter any architectural competitions that the UIA or their member sections have declared to be unacceptable.

5.7 Standard: Architects shall not when appointed as competition assessors subsequently act in any other capacity for the work.

5.8 Standard: Architects shall not maliciously or unfairly criticise or attempt to discredit another architect's work.

5.9 Standard: The architect shall, on being approached to undertake a project or other professional work upon which he/she knows or can ascertain by reasonable inquiry that another architect has a current appointment with the same client for the same project or professional work, notify the other architect.

5.10 Standard: Architects shall, when appointed to give an opinion on the work of another architect, notify the other architect, unless it can be shown to be prejudicial to prospective or actual litigation to do so.

5.11 Standard: Architects shall provide their associates and employees with a suitable working environment, compensate them fairly, and facilitate their professional development.

5.12 Standard: Architects shall ensure that their personal and professional finances are managed legally and prudently.

5.13 Standard: Architects shall build their professional reputation on the merits of their own service and performance and should recognise and give credit to others for professional work performed.

International Union of Architects

Recommended Guidelines for the UIA Accord On Recommended International Standards of Professionalism in Architectural Practice Policy on Continuing Professional Development

October 31, 1997
Revised March 11, 1998
Revised April 17, 1998
Revised December 10—12, 1998
Adopted July, 1999

∽∽∽∽∽∽∽∽∽∽∽∽∽∽∽∽∽∽∽∽∽∽∽∽∽∽∽∽∽∽

UIA Professional Practice Program Joint Secretariat

The American Institute of Architects	The Architectural Society of China
Co-Director James A. Scheeler, FAIA	Co-Director Zhang Qinnan, Vice President
1735 New York Avenue, NW	Bai Wan Zhuang, West District
Washington, DC 20006	Beijing, China 100835
Telephone: 202 - 6267315	Telephone: 8610 - 88082239
Facsimile: 202 - 6267421	Facsimile: 8610 - 88082222

Accord Policy on Continuing Professional Development

The UIA encourages its member sections to advocate continuing professional development as a duty of membership in the public interest. Architects must be sure they are capable of providing the services they offer, and codes of conduct must oblige architects to maintain a known standard in a variety of areas described under the "Fundamental Requirements of an Architect" and in future variations thereof. In the meantime, the UIA must monitor the development in continuing professional development for registration renewal, recommend guidelines among all nations to facilitate reciprocity, and continue to develop policy on this subject.

Recommended Guidelines for the Accord Policy on Continuing Professional Development

Continuing professional development does not refer to formal education leading to a more advanced degree, but to a life-long learning process that maintains, enhances, or increases the knowledge and skills of architects to ensure their knowledge and ability relevant to the needs of society.

The policy of the UIA encourages its member sections to advocate continuing professional development as the responsibility of each individual architect. Continuing professional development for architects is also in the public interest.

The UIA continuing professional development guidelines are intended to provide UIA member sections with a set of standards by which they can judge their existing professional development policies. This will ensure compatibility of polices and will provide for reciprocity and portability of professional development credits across member sections in the future.

One of the initial goals of the UIA continuing professional development guidelines will be to provide a framework for interrecognition of continuing professional development credits among the UIA member sections.

Key elements of a UIA member section continuing professional development system should include:

- Recommended procedures for identifying, screening, and evaluating continuing professional development services and courses;
- Recommended program criteria covering both self-study programs and registered continuing professional development providers' programs;
- Recommended criteria for incorporating research and needs assessments into the design and delivery of continuing professional development programs;
- Recommended procedures to assure that emphasis of the learning is placed on the learner and knowledge gained, including incentives for learning activities that increase interaction between the participant and the provider, e.g., interactive programs may give more credit for the same amount of time spent than that given for noninteractive programs;
- Recommended program quality levels and standards as a tool to assess the actual learning that occurs during a program and as an incentive to providers and participants to increase interaction that takes place and to involve the participants in the learning; participants should earn credits based on the educational quality of a program as well as the length of the program (seat time);
- Recommended procedures by which providers will give feedback to users and collect course evaluations to monitor the effectiveness of the activity
- A recommended record-keeping system that is timely and accurate for both providers and users that provides proven portability of continuing professional development credits and reporting to permit worldwide program availability to architects by registered providers and a credible basis for meeting the demands of registration agencies and professional societies that require continuing professional development credits for maintaining registration or membership;
- Recommendations for requirements of number of learning units to be earned each calendar year;
- Recommendations for minimum requirements of continuing professional development credits in subjects related to the protection of public health, safety, and welfare.

The UIA continuing professional development system should set high-quality educational standards for participating UIA member sections with a large number of registered providers contributing their knowledge, skills, and research to a successful endeavor.

International Union of Architects

Recommended Guidelines for the UIA Accord On Recommended International Standards of Professionalism in Architectural Practice Policy on Practice in a Host Nation

As approved by Council during the 95th Session held in Barcelona, Spain, 26-28 February 2002

UIA Professional Practice Program Joint Secretariat

The American Institute of Architects
Co-Director James A. Scheeler, FAIA
1735 New York Avenue, NW
Washington, DC 20006
Telephone: 202 – 6267315
Facsimile: 202 – 6267421

The Architectural Society of China

Co-Director Prof. Xu Anzhi
College of Architecture and Civil Engineering
Shenzhen University
Shenzhen, China 518060
Telephone: 8610 – 88082239
Facsimile: 8610 – 88082222

Accord Policy on Practice in a Host Nation

Architects providing architectural services on a project in a country in which they are not registered shall collaborate with a local architect to ensure that proper and effective understanding is given to legal, environmental, social, cultural, and heritage factors. The conditions of the association should be determined by the parties alone in accordance with UIA ethical standards and local statutes and laws.

Recommended Guidelines for the Accord Policy on Practice in a Host Nation

Preamble

The UIA is committed to the encouragement of bilateral and multilateral recognition agreements within the context of the General Agreement on Tariffs and Trade (GATT), The General Agreements on Trade in Services (GATTS) and the World Trade Organization (WTO). Experience in the development of mutual recognition and/or free trade agreements for the architectural profession suggests that the process requires identification of gaps between the elements of the foreign professional qualifications standards and those of the local qualification standards, and the negotiation of means to bridge these gaps through establishment of equivalencies. It is a process that must recognize the sovereign right of each jurisdiction to establish its professional standards at whatever level it deems appropriate for the environmental, social, cultural, public health, safety, and welfare interests of its citizens.

The Accord acknowledges that there are differences in the standards, practices and conditions reflecting the diversity of cultures of the countries of UIA member sections and that the Accord represents a first step in an effort by representatives of the international community of architects to reach consensus on standards and practices that will best serve community interests. The UIA recognizes that bilateral and multilateral mutual recognition and/or free trade agreements may take time to negotiate and bring into operation, and therefore there is a need to provide sector specific guidelines and protocols for conditions where mutual recognition and/or free trade agreements do not yet exist.

The Accord Policy on Practice in a Host Nation is intended to include equal standing between the associated architects and provide a bridge to the time when mutual recognition and/or free trade agreements are prevalent rather than a rarity, as is now the case. The following guideline suggests provisions for a protocol recommended for adoption by UIA member sections seeking to provide an appropriate mechanism for recognizing practice in a host nation by a foreign architect.

Introduction

In most jurisdictions architects must be registered, licensed, or certified in order to practice architecture. Practice in a Host Nation covers the situation when individual architects or corporate entities of architects have been commissioned to design projects in a country in which they are not registered, licensed, or certified.

The UIA recognizes the need for the responsible mobility of architects and their ability to provide services in foreign jurisdictions. It is the goal of the UIA that an architect recognized by the relevant authority of the nation/state in any UIA member section should be recognized as being able to be registered, licensed, or certified through bilateral or multilateral agreements and be able to establish in those nation/states as an architect by the relevant authorities in the nations/states of all UIA member sections.

The UIA also recognizes a need to promote the awareness of local environmental, social, and cultural factors and ethical and legal standards. To this end, the UIA Assembly has approved the Second Edition of the UIA Accord on Recommended International Standards of Professionalism in Architectural Practice. While the Accord and related Policy Guidelines are intended to define best practice for the architectural profession and the standards to which the profession aspires, they are also intended to make it easier for interested parties to negotiate mutual recognition and/or free trade agreements allowing portability of architectural credentials and/or services.

The long established UIA International Code of Ethics on Consulting Services requires that "every consultant from a foreign country...shall associate and work harmoniously with consultants or professionals of the country where the project is located."

Guidelines for Practice in a Host Nation

It is recommended that the member sections or nation/states of the International Union of Architects adopting this guideline agree that the UIA Accord on Recommended International Standards of Professionalism in Architectural Practice (referred hereafter as the Accord) establishes a policy framework for the negotiation of agreements under which local and foreign architects collaborate.

While the UIA Accord and related guidelines attempt to establish standards for the international practice of architecture, it is recognized that there are differences in the traditions and practices of the UIA member section countries.

Architects entering into 'Practice in a Host Nation' agreements should agree 1) that arrangements affecting professional liability, insurance, the jurisdiction of the courts, and similar issues are covered by local statutes or considered business arrangements and most appropriately negotiated by the local architect, the foreign architect, and the client and should be formalized in the agreements between and among the parties; 2) that public liability, statutes, and laws affecting the conduct of the architect and the practice of architecture are matters to be appraised by and the responsibility of both architects; and 3) that the following conditions shall apply for the practice of architecture by foreign architects in local jurisdictions:

1. In this guideline an architect is a professional recognized and registered/licensed/certified by a relevant authority in a nation/state. A local architect is the entity registered/licensed/certified and practicing in the nation where the project is located. A foreign architect is the entity registered/licensed/certified and practicing in a jurisdiction/country but is not registered/licensed/certified in the jurisdiction where the project is located.

2. Where there is no mutual recognition or free trade agreement between the relevant authorities of the host country and that of the foreign architects' country:

- Foreign architects registered/licensed/certified by a relevant authority in their own countries but not in the host country should be admitted individually and permitted to practice in association with registered/licensed/certified local architects, in accordance with local laws and practices.
- Foreign architects coming from nation/states that do not have relevant authorities dealing with issues of registration/licensing/certification should be required to be subjected to the registration/licensing/certification standards in force in the nation/states where the projects are located.

- A foreign architect should not be permitted to enter into an arrangement to provide services in another jurisdiction without the material participation of a local architect from that jurisdiction. This provision should not apply to the submission of entries in international competitions by eligible foreign architects. Should that foreign architect's competition submittal be selected, the foreign architect should be required to associate with a registered/licensed/certified local architect.

2.1 Foreign architects should:

a. Be prepared to demonstrate to the national or international relevant authority that they hold a current registration/license/certification from a relevant authority of a jurisdiction, which allows them to use the title "architect" and to engage in the unlimited practice of architecture in that jurisdiction.
b. Provide proof of their qualifications.
c. Certify that they are not subject to any criminal/ethical conviction.

2.2 Promptly after being selected as architect for a project in which a foreign architect is to be involved, the local architect should be required to provide a document to the relevant authority establishing the local architect's relationship with the foreign architects to include their qualifications, and a description of the specific project for which this arrangement has been made.

2.3 Foreign and local architects should make sure that in their collaboration they both have the necessary expertise and experience to meet the needs of the project.

2.4 Professional services rendered by the associated architects should jointly and severally be rendered by the foreign and qualified local architect involved in the project.

2.5 In any documents and when claiming credit for the project, the local architect and the foreign architect shall accurately represent their respective responsibilities on the project.

2.6 Upon request from a local relevant authority, it is expected that the foreign relevant authority of the UIA member section will agree to confirm the status of the foreign architect as suggested in paragraph 2.1.

2.7 A foreign architect should be required to agree to follow the laws, codes of ethics and conduct, building codes, etc. of the local jurisdiction.

2.8 Foreign and local architects forming collaborations should be required to enter into formal, fair and equitable agreements that uphold the ethical standards of the UIA. Numerous model documents and books have been developed that cover the principles involved and issues to be considered when developing an agreement between collaborating architects.

3. Where a mutual recognition agreement exists between the relevant authorities of two countries, this guideline is not relevant.

Practice in a Host Nation Drafting Panel
James A. Scheeler, Chairman
Luis M. Rossi
Carlos Maximiliano B. Fayet
Andreas Gottlieb Hempel
Tillman Prinz
Dato' Hahi Esa Bin Mohamed
Artur Jasinski
Edward D'Silva
Susan M. Allen

International Union of Architects

Recommended Guidelines for the UIA Accord On Recommended International Standards of Professionalism in Architectural Practice Policy on Intellectual Property and Copyright

As approved by Council during the 95th Session held in Barcelona, Spain, 26-28 February 2002

∽∽∽∽∽∽∽∽∽∽∽∽∽∽∽∽∽∽∽∽∽∽∽∽∽∽∽∽∽∽

UIA Professional Practice Program Joint Secretariat

The American Institute of Architects
Co-Director James A. Scheeler, FAIA
1735 New York Avenue, NW
Washington, DC 20006
Telephone: 202 – 6267315
Facsimile: 202 – 6267421

The Architectural Society of China
Co-Director Prof. Xu Anzhi
College of Architecture and Civil Engineering
Shenzhen University
Shenzhen, China 518060
Telephone: 8610 – 88082239
Facsimile: 8610 – 88082222

RECOMMENDED GUIDELINES FOR ACCORD POLICY ON INTELLECTUAL PROPERTY AND COPYRIGHT

Accord Policy on Intellectual Property and Copyright

That the national law of a UIA member section should entitle an architect to practice his/her profession without detriment to his/her authority and responsibility, and to retain ownership of the intellectual property and copyright of his/her work.

The UIA desires to develop and maintain the protection of the intellectual property and copyright of architects in their works in a manner as effective and uniform as possible. The UIA recognises the profound impact of the development and convergence of information and communication technologies on the creation and use of artistic works. Moreover the UIA emphasises the outstanding significance of copyright protection as an incentive to artistic creation, while recognising the need to maintain a balance between the rights of authors and the larger public interest.

The works mentioned in this Guideline should enjoy protection in all countries of the Union. This protection shall operate for the benefit of the author and his/her successors in title.

Preamble

Architectural services are characterised by the intellectual capacity of the architect. Qualified architects are capable of developing creative building solutions by-applying their knowledge and skill to a consideration of the client's requirements. The ideas and concepts created by the intellectual activity of architects are the products, which enables them to practice as architects. The importance of this creative intellectual endeavour requires that it have strong protection. Protection of-intellectual property rights guarantee that others cannot exploit the intellectual effort of architects and other creators of original works. This protection provides an incentive for further creative and innovative development in architecture for the benefit of climents and the public. This fact is essential for cultural improvement in architecture, which becomes more and more important in a globalized world in order to allow people to identify themselves with their built environment.

Intellectual Property means the rights resulting from intellectual activity in the industrial, scientific, literary or artistic fields, such as copyright rights, patent rights and others. As opposed to property of things, intellectual property covers the intellectual effort. Intellectual property is a general term made further concrete in patent law, copyright law or trade mark law.

Copyright covers specifically the intellectual effort regarding an artistic creation as opposed to the intellectual effort regarding the development of a pure technical process or object, which can only be protected under patent law. Copyright protects the interest of authors in relation to their creations and grants them the exclusive right of using their creations.

To provide services successfully in foreign countries architects must have the assurance that their intellectual property in their creative works are protected. Therefore it is important that intellectual property is protected in all countries to provide the necessary security for architects to confidently provide the best service to their clients. In this aspect the UIA Guideline on Practice in a Host Nation plays an important role in furthering cross-border services while protecting the intellectual property of architects.

Guideline

The following Guideline attempts to define the key concepts, issues and common eventualities relating to intellectual property rights in the field of architecture. These include definitions of "author" and the "works" that are subject to copyright and issues such as moral rights, protection, ownership and enforcement.

1. Author
1.1 Initial Ownership
The author of a work is the person who created that work, and is presumptively the owner of the copyright in the work. An author is always a natural person. Corporations, businesses or public entities may, however, own copyright of a work when the work is made under a consultancy agreements (See 1.2) or if the authors agree to assign their rights, provided that this is regulated in the relevant national

copyright law.

1.2 Works made under employment and consultant agreements
In the case of works authored by an architect while under an employment arrangement, the employer is presumptively the owner of the copyright, however this may be altered by explicit terms in the agreement of employment. In the case of works authored by an architect engaged as consultant, the architect is presumptively the owner of the copyright. However, if the law permits copyright may be transferred by agreement and with the consent of the author As copyright is a commercial property it is appropriate that payment, in addition to the fees paid for architectural services, is made to architects who agree to transferring the ownership of the copyright in their work.

1.3 Collective Works
Copyright in each separate contribution to a collective work is distinct from copyright in the collective work as a whole, and vests initially in the author of the contribution. In the absence of an express transfer of the copyright or of any rights under it, the owner of copyright in the collective work is presumed to have acquired only the privilege of reproducing and distributing the contribution as part of that particular collective work, any revision of that collective work, and any later collective work in the same series.

Copyright in a collective contribution in creating work vests in all authors of the collective work. All authors have the same rights connected to the copyright. Those rights can only be used collectively regarding the work as a whole.

2. Protected works
Copyright protects "original works of authorship" that are fixed in a tangible form of expression. The fixation need not be directly perceptible so long as it can be communicated with the aid of a machine or device. No publication or registration or other action is required to secure copyright. Copyright is secured automatically when the work is created for the first time.

2.1 Architectural works
Protected works under this guideline are original architectural works of authorship fixed in any tangible medium of expression, which represent a personal and original intellectual creation. Novelty, ingenuity, or aesthetic merit are not qualifying

criteria for protected works. The necessary standard of originality requires only that the work owes its existence to the efforts of the author and is not merely a copy of a pre-existing work.

Copyright protection extends only to documented or built works and not to ideas, procedures, methods of operation or mathematical concepts as such, because these works might be protected under the relevant technical protection rights such as patent rights. Copyright protection can cover any kind of architectural work.

2.2 Protection of specific architectural works

2.2.1 Documentation of architectural designs

Documentation of architectural designs and buildings either in electronic or hard-copy form can be protected under copyright. Apart from the plans, drawings, schemes, etc. of the object as a copyright protected work, the object itself can be protected, if realized in three dimensions. This refers also to works of town planning and urban design.

2.2.2 Expert opinions, specifications and other documents

The given form of representation of expert opinions, specifications and other documents can be protected under copyright, if they represent a personal creation. The protection does not cover the content of the document but only the form of representation, in order to distinguish copyright from technical protection rights such as patent right.

2.2.3 Buildings

A building can be protected under copyright, assuming the design fulfils the requirements of a personal creation of originality (See above 2.1). Similarly, parts of buildings or the combination of buildings can be protected as well as the newly creative assembled combination of already known elements as an ensemble. Style, taste, aesthetic value or fashion are of no importance to the question, if the work has the necessary creative character. Any building or architectural work can be protected under copyright.

3. Protection of the author's interests

The architect as owner of copyright has the exclusive rights to authorise the reproduction of her/his works in copies, as long as they are protected under copyright law. Unauthorised reproduction by others may give the right to the author to initiate legal action.

However it must be noted that in the field of architecture many building elements are already known, such as doors, windows, roofs and walls and hence limit the architectural creation. This is why the issue of infringement of copyright only arises, if a unique concept of a building, an extravagant technical detail or an extraordinary appearance of a building is copied. The mere influence of existing architectural works on the design of new-architectural works does not constitute an infringement of copyright.

Droit Moral/ Moral Rights
The so-called "moral rights" include the right of attribution and the right of integrity in the protected works. These rights provide for recognition of authorship and protection of the works against violations.

3.1 Right to publish architectural work

Subject to the laws relating to privacy authors of architectural works should enjoy the exclusive right to publish their work. This right applies to architects only in a limited way because in general they have no say in the-publication of the building they design for the client. However, the architect should have the power to decide if, when and how plans and pictures of his work are going to be published. In addition designs entered in architectural competitions should only be published and exhibited in accordance with the conditions of the competition or with the specific approval of the author.

Works produced by students in the course of their studies should only be published and exhibited as required for the purpose of scholarly evaluation and criticism. Students work should not be published for other reasons without the consent of the student.

The use of a copyrighted work for purposes such as criticism, comment, news reporting, teaching, scholarship or research should not be an infringement of copyright. The use for such purposes is sometimes referred as the 'fair use' exception to copyright protection. There are reasonable limitations on such fair use; which has the effect of reducing the author's market for the copyrighted work.

3.2 Recognition of authorship

Authors should have the right to put their name on their work and to have the work attributed to them when it is published. This recognition is especially important for sketches, plans and other documents, but it applies also to the built facility. Even though a copyright notice is not a condition of copyright protection, the notice should be displayed, particularly on drawings and other architectural documents, for clarification. In doing so, the author can avoid a defence of innocent infringement in mitigation of actual or statutory damages. A possible copyright notice could read: "Copyright© Associated Architects, Inc. 1999".

3.3 Violation of architectural works

Besides the author's economic rights and even if those rights are transferred someone else the author should have the right to claim authorship of the work and to object to plagiarism, distortion, mutilation or other modification of, or other derogatory action in relation to, the said work, which would be prejudicial to his honour or reputation. This right, sometimes described as the Moral Right to Integrity in created work, should be maintained even after the death of the author, at least until the expiry of the economic rights, and should be exercisable by the persons or institutions authorised by the legislation of the country where protection is claimed.

3.4 Alterations-Balancing the interests of owners and architects when buildings are altered

The long duration of a building makes it probable that adaptations, extensions or other changes are necessary. The client invested once in that building and must have the possibility to alter it according to his economical needs. The owner or user of the building must have the right to adapt the building to changing needs or purposes, which often includes changes in the architectural concept. Also changing public building regulations may require the alteration of a building.

At the same time the reputation of architects is largely established by their built works. Alterations to a building therefore have the potential to denigrate the reputation of the architect who is publicly known to be the author of the building. The alteration must ensure that the architects' personal interest for consistency of the building is safeguarded and their architectural capabilities are not disparaged or devalued in public. The architect is endangered that after unauthorized alteration of his publicized work he is still recognized as the architect of that work. In the eyes of the public the author will then be seen as having made those new architectural expressions, which can damage his/her image.

It is necessary, therefore, to find a balance between the two interests: Interest of the owner for alteration and the interest of the author for consistency. In finding that balance one has to take into account that the original architect of a building has more insight than anybody else regarding possibilities in developing the building for changing demands. Because of the original architect's profound knowledge about the design, construction and environment of his creation only that architect is in the best position to develop the necessary solutions while respecting the design of the building and its aesthetic qualities. Consequently it is recommended that the author of a building should have the legal right to be consulted prior to the building being altered. Such a right would not prohibit owners from proceeding with alterations to suit their requirements. However a consultation conducted in good faith provides an opportunity for the owner to consider maintaining the integrity of the original design or, if necessary, for the architect to publicly dissociate from the altered works.

3.5 Destruction
The right of objecting to the alteration of the work should include also the right to object to its destruction. Contrary to alteration the destruction of the work does not include the danger that the architect will be recognized as the author of the altered work. However the destruction of the work still violates the moral right of integrity in the architects work. The author has an interest in the work continuing to establish and maintain his or her professional reputation. Therefore the right of the owner to demolish the building conflicts with the moral right of the architect. Hence the right for destruction of a building must be balanced with the author's right for consistency.

4. Term of Protection
The term of copyright protection should extend to fifty years beyond the death of the author of the work.

5. Enforcement of Copyrights
The UIA recommends that laws and enforcement procedures are available that permit effective action against any act of infringement of intellectual property rights covered by this Guideline. These procedures should be applied in such a manner as to avoid the creation of barriers to legitimate trade and to provide for safeguards

against their abuse. Procedures concerning the enforcement of intellectual property rights should be fair and equitable. They should not be unnecessarily complicated or costly, or entail unreasonable time limits or unwarranted delays.

6. Ownership of architectural plans

The legal classification of ownership of architectural plans differs between the common-law countries and the countries under the Napoleonic Code. In common-law countries the architect's documents are normally treated as instruments of service and the architect retains the ownership while the client enjoys the license, by contract, to use the documents to build the building. In countries under the Napoleonic Code the architect's documents become property of the client after termination of the contract. The architect is obliged by contract to hand out the documents to the client. This situation has an effect on the intellectual property: Whereas in the common-law countries the architects are both the proprietor of the actual and intellectual property of their documents, the architects in countries under the Napoleonic Code are only proprietor of the intellectual property of the documents, the proprietor of the actual document is the client.

7. Co-operation between UIA member sections

Each UIA member section shall enter, upon request, into consultations with any other UIA member section which has cause to believe that an intellectual property right owner, who must be a national or domiciliary of the UIA member section being addressed for consultations, is undertaking practices in violation of the requesting UIA member section's laws and regulations on the subject matter of this Guideline.

The UIA member section addressed shall accord full and sympathetic consideration to, and shall afford adequate opportunity for, consultations with the requesting UIA member section. It shall, furthermore, co-operate through supply of publicly available non-confidential relevant information.

8. Damages

Jurisdictions should have the authority to order a person who has knowingly infringed copyright to pay adequate compensation to the copyright holder.

9. Institutional Arrangements; Final Provisions

The UIA shall monitor the operation of this Guideline, and in particular UIA member sections' compliance with their obligations hereunder, and shall afford UIA member sections the opportunity of consulting on matters relating to the aspects of intellectual property rights. It shall carry out such other responsibilities as assigned to it by the UIA member sections; and it shall, in particular, provide any assistance requested by them in the context of dispute settlement procedures. UIA member sections agree to co-operate with each other with a view to eliminating international trade in goods infringing intellectual property rights.

November 2001